Electrical Engineering: Fundamentals

Electrical Engineering: Fundamentals

Hannah Clark

STATES
ACADEMIC PRESS
www.statesacademicpress.com

Published by States Academic Press,
109 South 5th Street,
Brooklyn, NY 11249, USA

ISBN: 978-1-63989-172-6

Cataloging-in-Publication Data

Electrical engineering : fundamentals / Hannah Clark.
p. cm.
Includes bibliographical references and index.
ISBN 978-1-63989-172-6
1. Electrical engineering. 2. Engineering. I. Clark, Hannah.
TK145 .E44 2022
621.3--dc23

For information on all States Academic Press publications
visit our website at www.statesacademicpress.com

STATES
ACADEMIC PRESS

Contents

Preface

The branch of engineering which deals with the study, design and application of devices that work on the principles of electricity, electronics and electromagnetism is called electrical engineering. This discipline can be further divided into various sub-fields such as radio frequency engineering, power engineering, instrumentation and control engineering, telecommunications engineering, and electronics engineering. It finds extensive application in the field of communication systems, radar and navigation systems, power generation and distribution, etc. A wide range of equipment and tools are used in modern day electrical engineering like oscilloscope, multi-meter, network analyzers and spectrum analyzers. This book attempts to understand the multiple branches that fall under the discipline of electrical engineering and how such concepts have practical applications. The topics included herein on electrical engineering are of utmost significance and bound to provide incredible insights to readers. Those with an interest in this field would find this book helpful.

A short introduction to every chapter is written below to provide an overview of the content of the book:

Chapter 1 - Electrical Engineering is a branch of engineering that deals with the study, design, application and maintenance of electrical control systems, machinery and equipment. It further branches out into energy and power engineering, telecommunication engineering, computer engineering and electronic engineering. This chapter introduces the reader to electrical engineering and its associated sub-disciplines; **Chapter 2 -** The stream of charged particles moving through space or an electrical conductor is known as current. The potential difference in electrical change between two points in an electrical field is known as voltage. This chapter closely examines the basic electrical quantities and their related aspects to provide a thorough understanding of the subject; **Chapter 3 -** An electrical circuit is a closed loop interconnection of electrical components consisting of electrical elements. The study of electrical circuits is done using circuit diagrams and phasor diagrams. Electrical impedance, network analysis, Ohm's Law, Kirchhoff's Circuit Laws, source transformation, Norton's theorem, compensation theorem, superposition theorem and Thevenin's theorem are a few topics discussed in detail in this chapter; **Chapter 4 -** An electric machine is a general term for machines using electromagnetic forces to convert electricity to mechanical power and vice versa. A few electric machines closely examined and explained in this chapter are the rotary converter, metadyne, Dc motor, Ac motor, induction motor, transformer and doubly-fed electric machine; **Chapter 5 -** In electrical engineering, an exact indication of an amount of object or entity is crucial for a successful experiment or design of a system. This chapter's main focus is on the various units of measurement employed in electrical engineering like volt, ampere, coulomb, ohm, farad and henry and they have been comprehensively covered to provide the reader with a clear understanding.

I extend my sincere thanks to the publisher for considering me worthy of this task. Finally, I thank my family for being a source of support and help.

Hannah Clark

1

Electrical Engineering: An Introduction

Electrical Engineering is a branch of engineering that deals with the study, design, application and maintenance of electrical control systems, machinery and equipment. It further branches out into energy and power engineering, telecommunication engineering, computer engineering and electronic engineering. This chapter introduces the reader to electrical engineering and its associated sub-disciplines.

Electrical engineering is one of the newer branches of engineering, and dates back to the late 19th century. It is the branch of engineering that deals with the technology of electricity. Electrical engineers work on a wide range of components, devices and systems, from tiny microchips to huge power station generators.

Early experiments with electricity included primitive batteries and static charges. However, the actual design, construction and manufacturing of useful devices and systems began with the implementation of Michael Faraday's Law of Induction, which essentially states that the voltage in a circuit is proportional to the rate of change in the magnetic field through the circuit. This law applies to the basic principles of the electric generator, the electric motor and the transformer. The advent of the modern age is marked by the introduction of electricity to homes, businesses and industry, all of which were made possible by electrical engineers.

Some of the most prominent pioneers in electrical engineering include Thomas Edison (electric light bulb), George Westinghouse (alternating current), Nikola Tesla (induction motor), Guglielmo Marconi (radio) and Philo T. Farnsworth (television). These innovators turned ideas and concepts about electricity into practical devices and systems that ushered in the modern age.

Since its early beginnings, the field of electrical engineering has grown and branched out into a number of specialized categories, including power generation and transmission systems, motors, batteries and control systems. Electrical engineering also includes electronics, which has itself branched into an even greater number of subcategories, such as radio frequency (RF) systems, telecommunications, remote sensing, signal processing, digital circuits, instrumentation, audio, video and optoelectronics.

The field of electronics was born with the invention of the thermionic valve diode vacuum tube in 1904 by John Ambrose Fleming. The vacuum tube basically acts as a current amplifier by outputting a multiple of its input current. It was the foundation of all electronics, including radios, television and radar, until the mid-20th century. It was largely supplanted by the transistor, which was developed in 1947 at AT&T's Bell

Laboratories by William Shockley, John Bardeen and Walter Brattain, for which they received the 1956 Nobel Prize in physics.

"Electrical engineers design, develops, test and supervise the manufacturing of electrical equipment, such as electric motors, radar and navigation systems, and communications systems and power generation equipment. "Electronics engineers design and develop electronic equipment, such as broadcast and communications systems" — from portable music players to global positioning systems (GPS).

If it's a practical, real-world device that produces, conducts or uses electricity, in all likelihood, it was designed by an electrical engineer. Additionally, engineers may conduct or write the specifications for destructive or non-destructive testing of the performance, reliability and long-term durability of devices and components.

Today's electrical engineers design electrical devices and systems using basic components such as conductors, coils, magnets, batteries, switches, resistors, capacitors, inductors, diodes and transistors. Nearly all electrical and electronic devices, from the generators at an electric power plant to the microprocessors in your phone, use these few basic components.

Critical skills needed in electrical engineering include an in-depth understanding of electrical and electronic theory, mathematics and materials. This knowledge allows engineers to design circuits to perform specific functions and meet requirements for safety, reliability and energy efficiency, and to predict how they will behave, before a hardware design is implemented. Sometimes, though, circuits are constructed on "breadboards," or prototype circuit boards made on computer numeric controlled (CNC) machines for testing before they are put into production.

Electrical engineers are increasingly relying on computer-aided design (CAD) systems to create schematics and lay out circuits. They also use computers to simulate how electrical devices and systems will function. Computer simulations can be used to model a national power grid or a microprocessor; therefore, proficiency with computers is essential for electrical engineers. In addition to speeding up the process of drafting schematics, printed circuit board (PCB) layouts and blueprints for electrical and electronic devices, CAD systems allow for quick and easy modifications of designs and rapid prototyping using CNC machines.

Energy and Power Engineering

Energy Engineering

Energy engineering is a broad field of engineering that involves energy management, energy efficiency, renewable energy, and environmental maintenance. Not only must energy engineers be familiar with renewable energy, they must also know about oil and natural-gas energies. Energy engineering is an interdisciplinary collaboration between all of the energy technologies.

Energy engineering is all about reducing the world's energy consumption. Energy is the building blocks of everything and is required to power or build anything that is used in the world. Engineering applied to the energy sector focuses on increasing the energy efficiency of companies, machinery and households.

Energy engineers achieve energy saving through clever building designs, better use of materials and through renewable energy devices. Engineers often conduct building audits and inspections to first identify where energy can be saved. They use light level meters, thermo-guns and data logging to measure the energy losses.

Human beings have been transferring energy from one form to another since their first use of fire. Energy and powered devices are an integral part of society. Humanity's earliest days saw the discovery of fire through wood combustion (burning of wood), and the use of charcoal for smelting metals dates back as early as 5000 BC. Powered devices using natural energy sources such as water and wind were introduced by the Ancient Greeks and were commonly used until the 18th century, when the steam engine revolutionized the way devices could be powered. Various natural oils were also used for a range of purposes, such as whale oil for lamps.

Figure: Energy was first discovered through the burning of wood to create heat.

The Industrial Revolution led to the massive use of coal as fuel, and the extraction of petroleum and various other oils became extremely important with the advent of internal combustion engines. Electrical power, also based on fossil fuels, became widespread at the end of the 19th century and the production of cleaner electrical energy through hydro-power, nuclear, geothermal, and solar means is a topic even more relevant to today's world.

The efficiency of the transfer of energy is a new field. The oil crisis of 1973 and the energy crisis of 1979 brought to light the need to get more work out of less energy.

Alternative energy is renewable energy that comes from natural sources that are constantly replenished, and which do not harm the environment.

Compared to non-renewable fossil fuels they have lower carbon emissions. Wind power, geothermal power, hydro-power, biomass power, and solar power are examples of alternative energy, or renewables.

Energy sources such as solar and wind have improved over the years in both efficiency and adoption by the public. For both industries, engineers will be needed for research and for adapting new technology in the design, installation, and maintenance of new wind turbines and solar panels.

Power Engineering

Power Engineering is one of the earliest fields that have developed within Electrical Engineering. It deals with generation, transmission and distribution of electric power. Power engineers also work on a variety of power devices and on power conversion (the process of transforming power from one form into another, as in electromechanical or electrochemical processes).

Many power engineers are part of the large team that builds, maintains and develops the large networks that connects power generators with users of this power ("the power grid"). These engineers, who work for power utility companies of governments that maintain power grids design components for the grid, architectures for the grid and devices that either supply power to the grid or draw power from it. Devices that power engineers design and work with include generators, transformers, circuit breakers, relays and transmission lines. Systems the power engineer works on include electrical substations (a subsidiary station of an electricity generation, transmission and distribution system where voltage is transformed from high to low or the reverse using transformers.) Some power engineers also work on smaller "off grid networks" that generate and supply electricity to independent plants or remote areas. A separate area of expertise is generation, transmission and distribution of power on stand-alone plants such as planes and ships.

Power engineering is often analysed along the three components of generation, transmission and distribution.

Power engineers that work on generation convert other forms of energy into electric power. These sources of power include fossil fuels such as coal and natural gas, hydropower, nuclear power, solar power, and wind power.

Power engineers that work on transmission are in charge of moving power from the power station where the power is generated to the location of the customer. Power engineers that work on distribution are developing and maintaining systems that distribute power to end user in voltages that can be used by the user's equipment.

Power engineers deal with devices (motors, batteries, capacitors); processes and phenomena (such as power conversion, power drop and blackouts); analysis and design (such as estimation of the stability of a power network and power flow studies); and areas such as renewable energy and environmentally-friendly power systems. In addition to power utilities, power companies and organizations that maintain power networks, some power engineers work for universities and research institutions that advance the state of the art in power engineering and educate the next generation of power engineers.

In spite of its long history, power engineering is a vibrant and challenging discipline. Power engineers are in charge of very large systems whose availability and reliability are critical to society's ability to function and develop. The increase in demand in power, environmental and economic constraints, and the scarcity of some sources of power (such as fossil fuels) pose significant challenges to modern power engineers. These require new processes and techniques, new devices, and integration of other disciplines (such as business and law) in the design and implementation process.

Telecommunications Engineering

Telecommunications engineering is a discipline founded around the exchange of information across channels via wired or wireless means. It brings together all of the elements of electrical engineering, including computer engineering and system engineering, to create and improve telecommunication systems.

Telecom engineers work to develop, design and maintain voice and data communications systems, which include fiber, satellite, wired and unwired, as well as the encoding, encryption and compression of data. Put simply, telecommunications engineering can be found in just about every aspect of our lives, from GPS navigation to the internet.

The work of telecommunications engineers range from creating basic circuit designs to deploying wireless networks. They are responsible for designing and overseeing the installation of telecommunications equipment and facilities, such as complex electronic switching systems, copper wire telephone facilities, fiber optics cabling or internet protocol data systems.

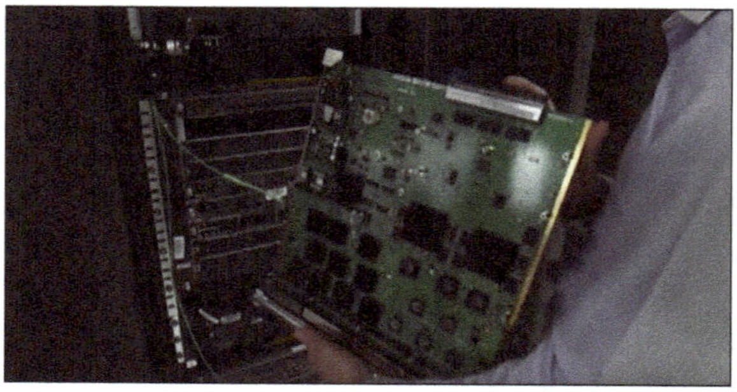

Some of the main areas of focus for telecommunications engineers are the installation of high-speed broadband computer networks, and optical and wireless or satellite communications.

Roles

Telecommunications engineers are part of every process of creating a telecom system, dealing with both software and hardware. Some roles a telecom engineer might take on:

- Design – Electronic components, software, products, or systems for commercial, industrial, medical, military or scientific applications.
- Develop – Maintenance and testing procedures for electronic components and equipment.
- Test – Evaluate systems and recommend design modifications or equipment repair.
- Debug – Inspect electronic equipment, instruments and systems to make sure they are safe.

With "5G" and the expansion of the "internet of things" ahead, telecommunications engineering will be as important as ever. The discipline will help expand both Low-Power Wide-Area Networks and networks that produce data speeds never before seen. Telecom engineers will have the important task of ensuring that telecommunications systems, from small components to entire networks, are running as effectively and efficiently as possible.

Computer Engineering

Computer engineering is the branch of engineering that integrates electronic engineering with computer sciences. Computer engineers design and develop computer systems and other technological devices.

Some of the most prominent pioneers in computer hardware include:

- Blaise Pascal, inventor of the mechanical calculator.
- Charles Babbage, inventor of the Analytical Engine.
- J. Presper Eckert and John Mauchly, builders of ENIAC, and later UNIVAC, the first computer to be sold commercially.
- Nicholas Metropolis, designer of the MANIAC I computer.
- William Shockley and his team at Bell Labs, inventors of the transistor.
- Seymour Cray, the "Father of Supercomputing".
- Steve Wozniak, designer of the Apple I and II personal computers and co-founder of Apple Computer along with Steve Jobs.
- Adam Osborne, developer of the first portable computer, the Osborne 1.
- Douglas Engelbart and Bill English, co-inventors of the computer mouse and developers of hypertext, computer networks and the graphical user interface (GUI).

Since those early days of computing, however, most significant advancements in computer hardware have been made by teams working at medium to large corporations. The early innovators in computer software include:

- George Boole, who formalized binary (Boolean) algebra.
- Alan Turing, who made enormous contributions to computer science, including inventing the Enigma code-breaking machine that broke the German ciphers in World War II and doing pioneering work in artificial intelligence and neural networks.
- John Backus, leader of the team that invented the first high-level programming language, Fortran (FORmula TRANslation).
- Dennis Ritchie and Ken Thompson, developers of the UNIX operating system and the C programming language.
- Gary Kildall, founder of Digital Research Inc. and inventor of the CP/M (Control Program for Microcomputers) operating system.
- Bill Gates, founder of Microsoft Corp.
- Vint Cerf and Bob Kahn, inventors of Transfer Control Protocol and Internet Protocol (TCP/IP) and sometimes referred to as the "Fathers of the Internet."

As with computer hardware, most significant advancements in computer software these days are made by companies rather than individuals.

Computer hardware engineers "research, design, develop and test computer systems and components such as processors, circuit boards, memory devices, networks and routers," according to the U.S. Bureau of Labor Statistics (BLS). Computer hardware includes:

- Microprocessors.
- Memory chips: Random-access memory (RAM), read-only memory (ROM) and non-volatile rewritable lash memory.
- Data storage devices: Hard disks, solid-state drives and optical drives.
- Input devices: Keyboards, mice, joysticks and gaming controllers, cameras, microphones, scanners, touch screens and remote sensors.
- Output devices: Printers, monitors, audio devices and remote controls.
- Networking components: Adapters, modems, switches and routers.

An important function of computer engineers is to integrate these components into computer and network systems. This all requires a good working knowledge of electrical engineering. Another important aspect of computer engineering is software development. Computer software includes:

- Operating systems.
- Applications: Word processing, spread sheets, accounting, database management, graphics, computer-assisted design (CAD), computer-assisted manufacturing (CAM), audio, video, media and games.
- Networking and communications: World Wide Web (WWW), voice over Internet Protocol (VOIP), instant messaging and email.
- Utilities: file handling, disk management, device drivers, archiving and backup systems.
- Programming languages: Editing, compiling and debugging.
- Security: Antivirus, irewalls, encryption and user authentication.

Electronic Engineering

Electronics engineering is a branch of electrical engineering focusing on the design and development of products using electronic technology.

Electronics engineering, sometimes called electronic engineering, is a modern engineering discipline focused on the design and development of products and systems using electronic technology. It developed in the early 20th century as electronic broadcasting methods, including radio and television, became more widespread. The use of electronic technologies in World War II, including sonar, radar, and advanced weaponry, further developed the field of electronics engineering.

Today, electronics engineering is a diverse field encompassing the design and development of items that use traditional analog technology and more high-tech digital components. Many of the devices we use every day are the work of electronics engineers, including smartphones, personal computers and Wi-Fi networks. It is the key force behind the growth of information technology in the modern age.

A variety of industries utilize electronics engineering to modernize their sector including telecommunications, radio and television, utilities, health care, science, personal technology manufacturing and government and military. The field of electronics engineering is as diverse as its applications, with many subsets including analog electronic engineering, radio frequency engineering, software engineering and systems engineering focusing on different applications of electronic technology.

Key Functions

Electronics engineers design, develop and oversee the production of electronic items and systems. Testing these products and their components is a key part of this process to make sure all electronics work efficiently, safely and reliably for personal and commercial consumers. The responsibilities of electronics engineers vary depending on their specialty and industry, but usually include:

- Planning electronics projects.
- Manufacturing electronics products.
- Testing and evaluating electronics products.
- Coordinating with relevant individuals.
- Writing usage recommendations.
- Performing maintenance and repairs.

Planning Electronics Projects

Electronics engineers are involved with the preliminary stages of any electronics product, planning how everything from small individual electronics to large-scale operations should be created. The planning process requires determining how each project should look and function, how much its development should cost, where budget money will be allocated and how long the project should take to complete. Electronics engineers must carefully consider the requirements of their employer or client and any international, national and local safety guidelines governing their work.

Manufacturing Electronic Products

Electronics engineers follow the plans developed by themselves or other electronics engineers to manufacture electronics products and systems. Most electronics engineers work as part of a team, creating individual electronics components or products, then assembling them with their colleagues' creations to make larger electronics works. This work requires an understanding and adherence to each project's specifications and international, national and local product and safety guidelines.

Testing and Evaluating Electrical Products

After assembly, electronics engineers conduct final tests before their electronics engineers are released. They assess whether their products operate as intended and adhere to the guidelines and specifications provided. Rigorous tests determine whether electronics products are durable and safe for consumers. Electronics engineers may also note any areas where their electronics items or components may be improved. Proposed changes are evaluated, to determine whether they add enough value to consumers to justify additional costs and time. If approved, the changes electronics engineers suggest are implemented before the product's release.

Coordinating with Relevant Individuals

Since they work as part of larger teams, electronics engineers spend a lot of time coordinating with others inside and outside their business. Collaborating with other electronics engineers ensures that their work benefits the projects they are working on. Brainstorming with other team members can enhance electronics projects. Depending on their seniority, electronics engineers may talk to vendors and suppliers, to source electronics components, and clients, to determine each project's requirements and provide status updates. Many electronics engineers also provide status updates to their project managers.

Electronics engineers may use various methods to coordinate their efforts with other individuals and businesses, including meetings in person, telephone calls, videoconferencing and email. They may also write and review written documents, including contracts and proposals.

Writing usage Recommendations

Electronics engineers write usage recommendations to ensure consumers use their products correctly. The recommendations electronics engineers write instruct users how to handle, operate and troubleshoot electronics products. They may also offer maintenance procedures. The instructions electronics engineers write help consumers use their products safely and efficiently, to maximize their lifespans.

Performing Maintenance and Repairs

Some electronics products, such as industrial machinery with electronics components, require regular maintenance from electronics engineers. If they detect any small problems during maintenance, they may perform repairs before these problems escalate or suggest replacing the goods. This often involves doing full or partial circuit traces to detect faults, then replacing components including drives, circuit boards or power supplies.

References

- Electrical-engineering-47571: livescience.com, Retrieved 06, August 2020
- What-is-energy-engineering: mechanicalengineeringhq.com, Retrieved 04, May 2020
- Engineering-basics, careers-in-power-energy: ee-scholarship.org, Retrieved 09, June 2020
- What-is-telecommunications-engineering, fundamentals- 20160720: rcrwireless.com, Retrieved 12, July 2020
- What-is-electronics-engineering, finding-a-job, career-advice: indeed.com, Retrieved 16, March 2020

2

Basic Electrical Quantities: Current, Voltage and Power

The stream of charged particles moving through space or an electrical conductor is known as current. The potential difference in electrical change between two points in an electrical field is known as voltage. This chapter closely examines the basic electrical quantities and their related aspects to provide a thorough understanding of the subject.

Electric Current

An electric current is a flow of electric charge in a circuit. More specifically, the electric current is the rate of charge flow past a given point in an electric circuit. The charge can be negatively charged electrons or positive charge carriers including protons, positive ions or holes.

The magnitude of the electric current is measured in coulombs per second, the common unit for this being the Ampere or amp which is designated by the letter 'A'. The Ampere or amp is widely used within electrical and electronic technology along with the multipliers like milliamp (0.001A), micro amp (0.000001A), and so forth. Current flow in a circuit is normally designated by the letter 'I', and this letter is used in equations like Ohms law where V=I·R.

The basic concept of current is that it is the movement of electrons within a substance. Electrons are minute particles that exist as part of the molecular structure of materials. Sometimes these electrons are held tightly within the molecules and other times they are held loosely and they are able to move around the structure relatively freely.

One very important point to note about the electrons is that they are charged particles - they carry a negative charge. If they move then an amount of charge moves and this is called current.

It is also worth noting that the number of electrons that able to move governs the ability of a particular substance to conduct electricity. Some materials allow current to move better than others.

The motion of the free electrons is normally very haphazard - it is random - as many electrons move in one direction as in another and as a result there is no overall movement of charge.

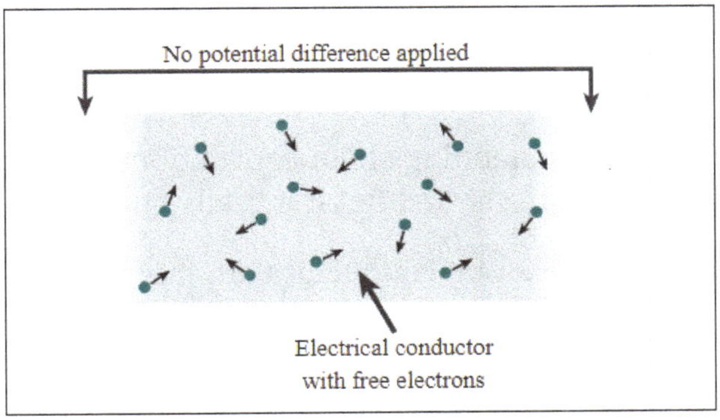

Figure: Random electron movement in a conductor with free electrons.

If a force acts on the electrons to move them in a particular direction, then they will all drift in the same direction, although still in a somewhat haphazard fashion, but there is an overall movement in one direction.

The force that acts on the electrons is called and electromotive force, or EMF, and its quantity is voltage measured in volts.

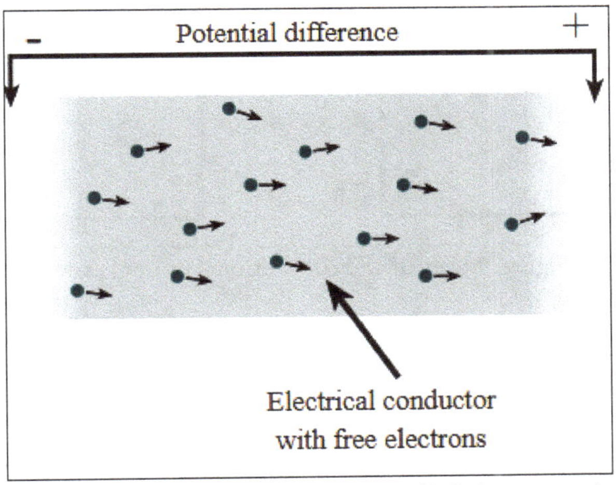

Figure: Electron flow under the action of applied electro-motive force.

To gain a little more understanding about what current is and how it acts in a conductor, it can be compared to water flow in a pipe. There are limitations to this comparison, but it serves as a very basic illustration of current and current flow.

The current can be considered to be like water flowing through a pipe. When pressure is placed on one end it forces the water to move in one direction and flow through the pipe. The amount of water flow is proportional to the pressure placed on the end. The pressure or force placed on the end can be likened to the electro-motive force.

When the pressure is applied to the pipe, or the water is allowed to flow as a result of a tap being opened, then the water flows virtually instantaneously. The same is true for the electrical current. To gain an idea of the flow of electrons, it takes 6.24 billion, billion electrons per second to flow for a current of one ampere.

Conventional Current and Electron Flow

There is often a lot of misunderstanding about conventional current flow and electron flow. This can be a little confusing at first but it is really quite straightforward.

The particles that carry charge along conductors are free electrons. The electric field direction within a circuit is by definition the direction that positive test charges are pushed. Thus, these negatively charged electrons move in the direction opposite the electric field.

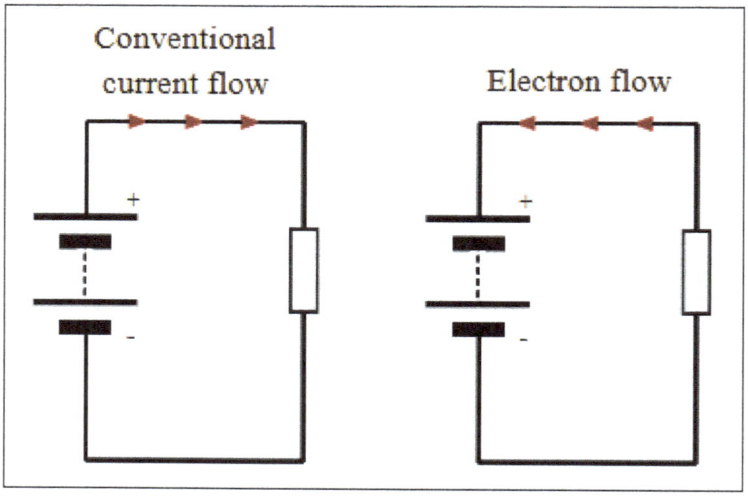

Figure: Electron and conventional current flow.

This came about because the initial investigations in static and dynamic electric currents was based upon what we would now call positive charge carriers. This meant that then early convention for the direction of an electric current was established as the direction that positive charges would move. This convention has remained and it is still used today.

- Conventional current flow: The conventional current flow is from positive to the negative terminal and indicates the direction that positive charges would flow.
- Electron flow: The electron flow is from negative to positive terminal. Electrons are negatively charged and are therefore attracted to the positive terminal as unlike charges attract.

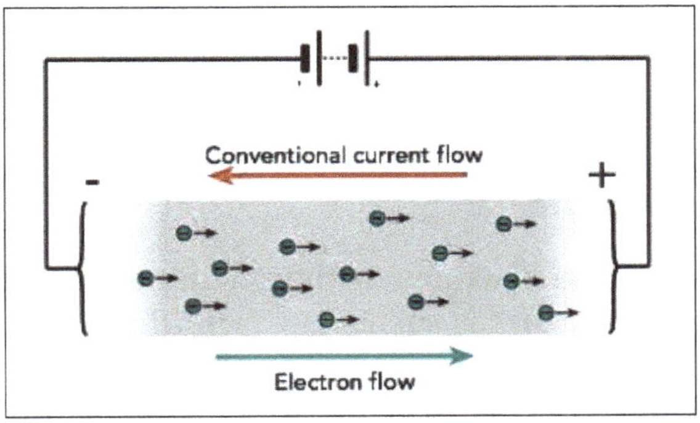

Figure: This is the convention that is used globally to this day, even though it may seem a little odd and out-dated.

Speed of Electron or Charge Movement

The speed of the transmission of electrical current is very different to that of the speed of the actual electron movement. The electron itself bounces around in the conductor, and possibly only makes progress along the conductor at the rate of a few millimetres a second. This means that in the case of alternating current, where the current changes direction 50 or 60 times per second, most of the electrons never make it out of the wire.

To take a different example, in the near-vacuum inside a cathode ray tube, the electrons travel in near-straight lines at about a tenth of the speed of light.

Effects of Current

When an electric current flows through a conductor there are a number of signs which tell that a current is flowing:

- Heat is dissipated: Possibly the most obvious is that heat is generated. If the current is small then the amount of heat generated is likely to be very small and may not be noticed. However if the current is larger than it is possible that a noticeable amount of heat is generated. An electric fire is a prime example showing how a current causes heat to be generated. The actual amount of heat is governed not only be the current, but also be the voltage and the resistance of the conductor.

- Magnetic effect: Another effect which can be noticed is that a magnetic field is built up around the conductor. If a current is flowing in conductor then it is possible to detect this. By placing a compass close to a wire carrying a reasonably large direct current, the compass needle can be seen to be deflected. Note this will not work with mains because the field is alternating too fast for the needle to respond and the two wires (live and neutral) close together in the same cable will cancel out the field.

The magnetic field generated by a current is put to good use in a number of areas. By winding a wire into a coil, the effect can be increased, and an electro-magnet can be made. Relays and a host of other items use the effect. Loudspeakers also use a varying current in a coil to cause vibrations to occur in a diaphragm which enable the electronic currents to be converted into sounds.

How to Measure Current

One important aspect of current is knowing the amount of current that may be flowing in a conductor. As electric current is such a key factor in electrical and electronic circuits, knowing what current is flowing is very important. There are many different ways is measuring current. One of the easiest is to use a multimeter. Although there are other methods of measuring current, this is the most common.

Current is one of the most important and fundamental elements within electrical and electronic technology. The current flowing in a circuit can be used in a variety of ways from generating heat to causing circuits to switch, or information to be stored in an integrated circuit.

Direct and Alternating Current

Alternating Current

Alternating current describes the flow of charge that changes direction periodically. As a result, the voltage level also reverses along with the current. AC is used to deliver power to houses, office buildings, etc.

Generating AC

AC can be produced using a device called an alternator. This device is a special type of electrical generator designed to produce alternating current.

A loop of wire is spun inside of a magnetic field, which induces a current along the wire. The rotation of the wire can come from any number of means: a wind turbine, a steam turbine, flowing water, and so on. Because the wire spins and enters a different magnetic polarity periodically, the voltage and current alternates on the wire.

To generate AC in a set of water pipes, we connect a mechanical crank to a piston that moves water in the pipes back and forth (our “alternating” current). Notice that the pinched section of pipe still provides resistance to the flow of water regardless of the direction of flow.

Waveforms

AC can come in a number of forms, as long as the voltage and current are alternating. If we hook up an oscilloscope to a circuit with AC and plot its voltage over time, we might see a number of different waveforms. The most common type of AC is the sine wave. The AC in most homes and offices has an oscillating voltage that produces a sine wave.

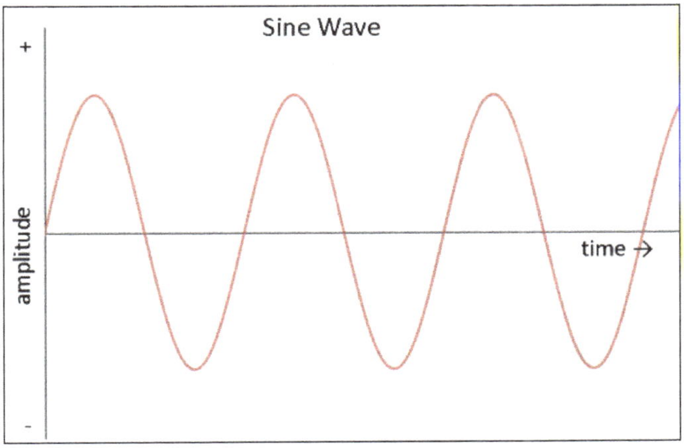

Other common forms of AC include the square wave and the triangle wave.

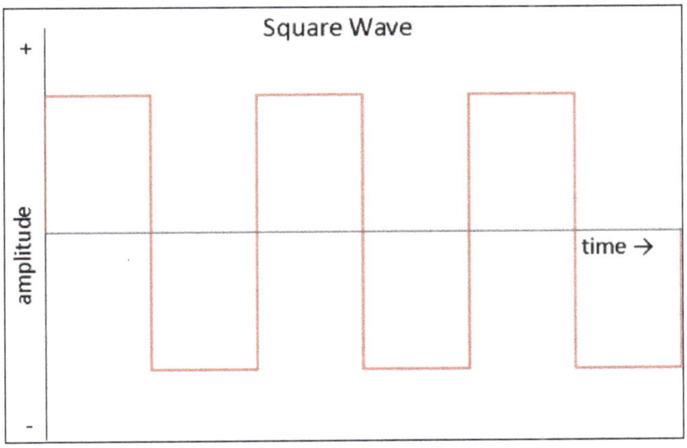

Square waves are often used in digital and switching electronics to test their operation.

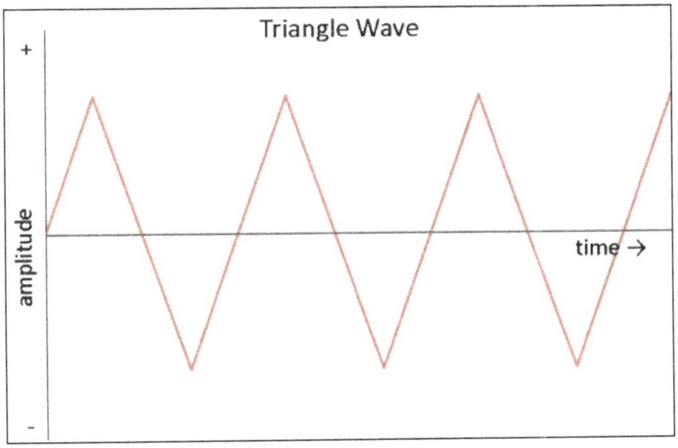

Triangle waves are found in sound synthesis and are useful for testing linear electronics like amplifiers.

Describing a Sine Wave

We often want to describe an AC waveform in mathematical terms. For this example, we will use the common sine wave. There are three parts to a sine wave: amplitude, frequency, and phase.

Looking at just voltage, we can describe a sine wave as the mathematical function:

$$V(t) = V_P \sin(2\pi ft + \phi)$$

V(t) is our voltage as a function of time, which means that our voltage changes as time changes. The equation to the right of the equals sign describes how the voltage changes over time.

V_P is the amplitude. This describes the maximum voltage that our sine wave can reach in direction, meaning that our voltage can be $+V_P$ volts, $-V_P$ volts, or somewhere in between.

The sin () function indicates that our voltage will be in the form of a periodic sine wave, which is a smooth oscillation around 0V.

2π is a constant that converts the freqency from cycles (in hertz) to angular frequnecy (radians per second).

f describes the frequency of the sine wave. This is given in the form of hertz or units per second. The frequency tells how many times a particular wave form (in this case, one cycle of our sine wave - a rise and a fall) occurs within one second.

t is our independent variable: time (measured in seconds). As time varies, our waveform varies.

φ describes the phase of the sine wave. Phase is a measure of how shifted the waveform is with respect to time. It is often given as a number between 0 and 360 and measured in degrees. Because of the periodic nature of the sine wave, if the wave form is shifted by 360° it becomes the same waveform again, as if it was shifted by 0°. For simplicity, we still assume that phase is 0° for the rest of this tutorial.

We can turn to our trusty outlet for a good example of how an AC waveform works. In the United States, the power provided to our homes is AC with about 170V zero-to-peak (amplitude) and 60Hz (frequency). We can plug these numbers into our formula to get the equation (remember that we are assuming our phase is 0):

$$V(t) = 170 \sin(2\pi 60 ft)$$

We can use our handy graphing calculator to graph this equation. If no graphing calculator is available we can use a free online graphing program like Desmos (Note that you might have to use 'y' instead of 'v' in the equation to see the graph).

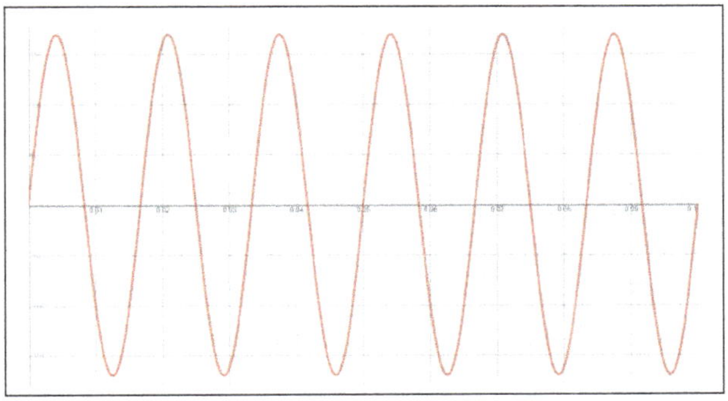

Notice that, as we predicted, the voltage rise up to 170V and down to -170V periodically. Additionally, 60 cycles of the sine wave occurs every second. If we were to measure the voltage in our outlets with an oscilloscope, this is what we would see.

You might have heard that AC voltage in the US is 120V. This is also correct. How? When talking about AC (since the voltage changes constantly), it is often easier to use an average or mean. To accomplish that, we use a method called "Root mean squared." (RMS). It is often helpful to use the RMS value for AC when you want to calculate electrical power. Even though, in our example, we had the voltage varying from -170V to 170V, the root mean square is 120V RMS.

Applications

Home and office outlets are almost always AC. This is because generating and transporting AC across long distances is relatively easy. At high voltages (over 110kV), less energy is lost in electrical power transmission. Higher voltages mean lower currents, and lower currents mean less heat generated in the power line due to resistance. AC can be converted to and from high voltages easily using transformers.

AC is also capable of powering electric motors. Motors and generators are the exact same device, but motors convert electrical energy into mechanical energy (if the shaft on a motor is spun, a voltage is generated at the terminals!). This is useful for many large appliances like dishwashers, refrigerators, and so on, which run on AC.

Direct Current

Direct current is a bit easier to understand than alternating current. Rather than oscillating back and forth, DC provides a constant voltage or current.

Generating DC

DC can be generated in a number of ways:

- An AC generator equipped with a device called a "commutator" can produce direct current.

- Use of a device called a “rectifier” that converts AC to DC.
- Batteries provide DC, which is generated from a chemical reaction inside of the battery.

DC is similar to a tank of water with a hose at the end. The tank can only push water one way: out the hose. Similar to our DC-producing battery, once the tank is empty, water no longer flows through the pipes.

Describing DC

DC is defined as the “unidirectional” flow of current; current only flows in one direction. Voltage and current can vary over time so long as the direction of flow does not change. To simplify things, we will assume that voltage is a constant. For example, we assume that a AA battery provides 1.5V, which can be described in mathematical terms as:

$$V(t) = 1.5V$$

If we plot this over time, we see a constant voltage:

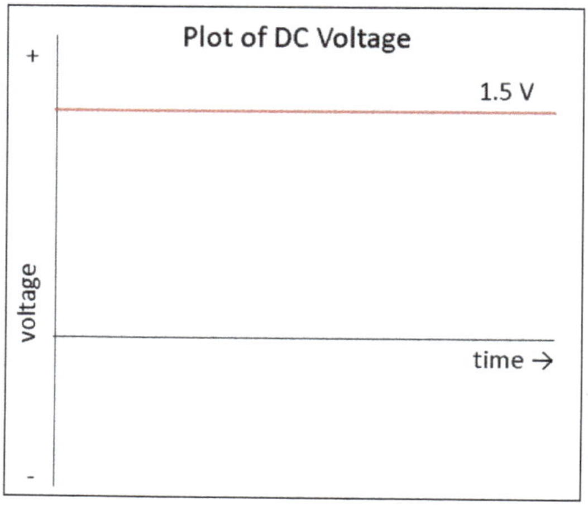

What does this mean? It means that we can count on most DC sources to provide a constant voltage over time. In reality, a battery will slowly lose its charge, meaning that the voltage will drop as the battery is used. For most purposes, we can assume that the voltage is constant.

Applications

Almost all electronics projects and parts for sale on SparkFun run on DC. Everything that runs off of a battery, plugs in to the wall with an AC adapter, or uses a USB cable for power relies on DC. Examples of DC electronics include:

- Cell phones.
- The LilyPad-based D&D Dice Gauntlet.
- Flat-screen TVs (AC goes into the TV, which is converted to DC).
- Flashlights.
- Hybrid and electric vehicles.

Battle of the Currents

Almost every home and business is wired for AC. However, this was not an overnight decision. In the late 1880s, a variety of inventions across the United States and Europe led to a full-scale battle between alternating current and direct current distribution.

In 1886, Ganz Works, an electric company located in Budapest, electrified all of Rome with AC. Thomas Edison, on the other hand, had constructed 121 DC power stations in the United States by 1887. A turning point in the battle came when George Westinghouse, a famous industrialist from Pittsburgh, purchased Nikola Tesla's patents for AC motors and transmission the next year.

AC vs. DC

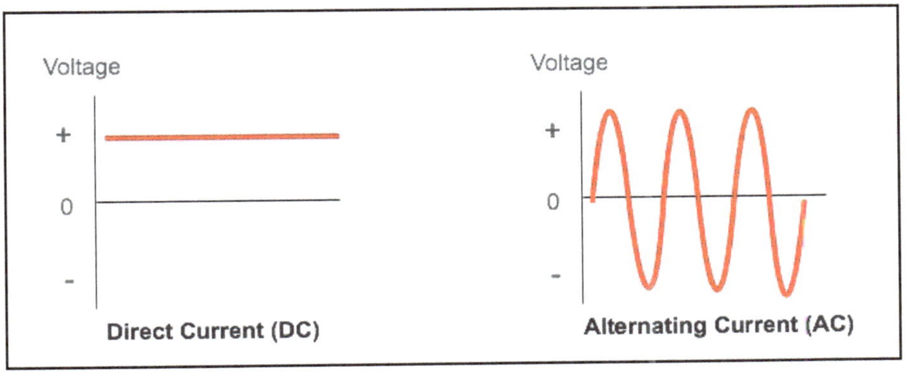

In the late 1800s, DC could not be easily converted to high voltages. As a result, Edison proposed a system of small, local power plants that would power individual neighborhoods or city sections. Power was distributed using three wires from the power plant: +110 volts, 0 volts, and -110 volts. Lights and motors could be connected between either the +110V or 110V socket and 0V (neutral). 110V allowed for some voltage drop between the plant and the load (home, office, etc.).

Even though the voltage drop across the power lines was accounted for, power plants needed to be located within 1 mile of the end user. This limitation made power distribution in rural areas extremely difficult, if not impossible.

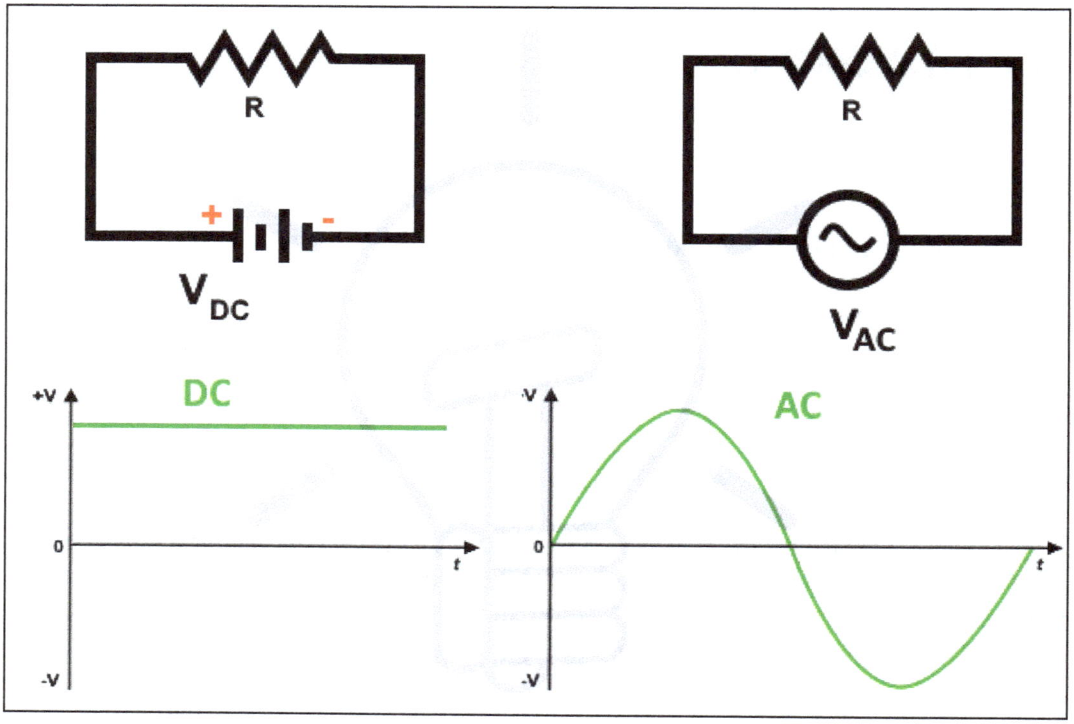

With Tesla's patents, Westinghouse worked to perfect the AC distribution system. Transformers provided an inexpensive method to step up the voltage of AC to several thousand volts and back down to usable levels. At higher voltages, the same power could be transmitted at much lower current, which meant less power lost due to resistance in the wires. As a result, large power plants could be located many miles away and service a greater number of people and buildings.

Edison's Smear Campaign

Over the next few years, Edison ran a campaign to highly discourage the use of AC in the United States, which included lobbying state legislatures and spreading disinformation about AC. Edison, also directed several technicians to publicly electrocute animals with AC in an attempt to show that AC was more dangerous than DC. In attempt to display these dangers, Harold P. Brown and Arthur Kennelly, employees of Edison, designed the first electric chair for the state of New York using AC.

The Rise of AC

In 1891, the International Electro-Technical Exhibition was held in Frankfurt, Germany and displayed the first long distance transmission of three-phase AC, which powered lights and motors at the exhibition. Several representatives from what would become General Electric were present and were subsequently impressed by the display. The following year, General Electric formed and began to invest in AC technology.

Figure: Edward Dean Adams Power Plant at Niagara Falls, 1896.

Westinghouse won a contract in 1893 to build a hydroelectric dam to harness the power of Niagara Falls and transmit AC to Buffalo, NY. The project was completed on November 16, 1896 and AC power began to power industries in Buffalo. This milestone marked the decline of DC in the United States. While Europe would adopt an AC standard of 220-240 volts at 50 Hz, the standard in North America would become 120 volts at 60 Hz.

High-Voltage Direct Current

Swiss engineer René Thury used a series of motor-generators to create a high-voltage DC system in the 1880s, which could be used to transmit DC power over long distances. However, due to the high cost and maintenance of the Thury systems, HVDC was never adopted for almost a century.

With the invention of semiconductor electronics in the 1970s, economically transforming between AC and DC became possible. Specialized equipment could be used to generate high voltage DC power (some reaching 800 kV). Parts of Europe have begun to employ HVDC lines to electrically connect various countries.

HVDC lines experience less loss than equivalent AC lines over extremely long distances. Additionally, HVDC allows different AC systems (e.g. 50 Hz and 60 Hz) to be connected. Despite its advantages, HVDC systems are more costly and less reliable than the common AC systems.

Resistive Heating

Whenever an electric current flows through a material that has some resistance (i.e., anything but a superconductor), it creates heat. This resistive heating is the result of "friction," as created by microscopic phenomena such as retarding forces and collisions involving the charge carriers (usually electrons); in formal terminology, the heat corresponds to the work done by the charge carriers in order to travel to a lower potential.

This heat generation may be intended by design, as in any heating appliance (for example, a toaster, an electric space heater, or an electric blanket). Such an appliance essentially consists of a conductor whose resistance is chosen so as to produce the desired amount of resistive heating. In other cases, resistive heating may be undesirable. Power lines are a classic example. For one, their purpose is to transmit energy, not to dissipate it; the energy converted to heat along the way is, in effect, lost (thus the term resistive losses). Furthermore, resistive heating of transmission and distribution lines is undesirable, since it causes thermal expansion of the conductors, making them sag. In extreme cases such as fault conditions, resistive heating can literally melt the wires.

Calculating Resistive Heating

There are two simple formulas for calculating the amount of heat dissipated in a resistor (i.e., any object with some resistance). This heat is measured in terms of power, which corresponds to energy per unit time. Thus, we are calculating a rate at which energy is being converted into heat inside a conductor. The first formula is:

$$P = I \times V$$

Where P is the power, I is the current through the resistor, and V is the voltage drop across the resistor.

Power is measured in units of watts (W), which correspond to amperes x volts. Thus, a current of one ampere flowing through a resistor across a voltage drop of one volt produces one watt of heat. Units of watts can also be expressed as joules per second. To conceptualize the magnitude of a watt, it helps to consider the heat created by a 100-watt light bulb, or a 1000-watt space heater.

The relationship P = I × V makes sense if we recall that voltage is a measure of energy per unit charge, while the current is the flow rate of charge. The product of current and voltage therefore tells us how many electrons are "passing through," multiplied by the amount of energy each electron loses in the form of heat as it goes, giving an overall rate of heat production. We can write this as:

$$\frac{Charge}{Time} \cdot \frac{Energy}{Charge} = \frac{Energy}{Time}$$

and see that, with the charge canceling out, units of current multiplied by units of voltage indeed give us units of power.

The second formula for calculating resistive heating is:

$$P = I^2 \times R$$

where P is the power, I is the current, and R is the resistance. This equation could be derived from the first one by substituting I . R for V (according to Ohm's law). This second formula is more frequently used in practice to calculate resistive heating, whereas the first formula has other, more general applications.

As we might infer from the equation, the units of watts also correspond to amperes^2 × ohms ($A^2 \times \Omega$). Thus, a current of one ampere flowing through a wire with one ohm resistance would heat this wire at a rate of one watt. Because the current is squared in the equation, two amperes through the same wire would heat it at a rate of 4 watts, and so on.

Example: A toaster oven draws a current of 6 A at a voltage of 120 V. It dissipates 720 W in the form of heat. We can see this in two ways:

First, using P= I × V, 120V × 6 A = 720 W. Alternatively, we could use the resistance, which is 20 V (20 V × 6 A = 120 V), and write $P = I^2 \times R$ / $(6A)^2 \times 20\ \Omega = 720$ W.

It is important to distinguish carefully how power depends on resistance, current, and voltage, since these are all interdependent. Obviously, the power dissipated will increase with increasing voltage and with increasing current. From the formula $P = I^2 \times R$, we might also expect power to increase with increasing resistance, assuming that the current remains constant. However, it may be incorrect to assume that we can vary resistance without varying the current.

Specifically, in many situations it is the voltage that remains (approximately) constant. For example, the voltage at a customer's wall outlet ideally remains at 120 V, regardless of how much power is consumed. The resistance is determined by the physical properties of the appliance: its intrinsic design, and, if applicable, a power setting (such as "high" or "low"). Given the standard voltage, then, the resistance determines the amount of current "drawn" by the appliance according to Ohm's law: higher resistance means lower current, and vice versa. In fact, resistance and current are inversely proportional in this case: if one doubles, the other is halved.

Effect of Resistance on Power Consumption

The key here is that resistive heating depends on the square of the current, meaning that the power is more sensitive to changes in current than resistance. Therefore, at constant voltage, the effect of a change in current outweighs the effect of the corresponding change in resistance. For example, decreasing the resistance (which, in and of itself, would tend to decrease resistive heating) causes the current to increase, which increases resistive heating by a greater factor. Thus, at constant voltage, the net effect of decreasing resistance is to increase power consumption. An appliance that draws more power has a lower internal resistance.

For an intuitive example, consider the extreme case of a short circuit, caused by an effectively zero resistance (usually unintentional). Suppose a thick metal bar were placed across the terminals of a car battery. A very large current would flow, the metal would become very hot, and the battery would be drawn down very rapidly. If a similar experiment were performed on a wall outlet by sticking, say, a fork into it, the high current would hopefully be interrupted by the circuit breaker before either the fork or the wires melted.

The other extreme case is simply an open circuit, where the two terminals are separate and the resistance of the air between them is infinite: here the current and the power consumption are obviously zero.

Example: Consider two incandescent light bulbs, with resistances of 240 V and 480 V. How much power do they each draw when connected to a 120 V outlet?

First we must compute the current through each bulb, using Ohm's law: Substituting V ¼ 120 V and R1 ¼ 240 V into $V = I \times R$, we obtain $I_1 = 0.5$ A. For $R_2 = 480$ V, we get $I_2 = 0.25$ A.

Now we can use these values for I and R in the power formula, $P = I^2 \times R$, which yields $P_1 = (0.5\ A)^2 \times 240\ V = 60\ W$ and $P_2 = (0.25\ A)^2 \times 480\ V = 30\ W$.

We see that at constant voltage, the bulb with twice the resistance draws half the power.

There are other situations, however, where the current rather than the voltage is constant. Transmission and distribution lines are an important case. Here, the reasoning suggested earlier does in fact apply, and resistive heating is directly proportional to resistance. The important difference between power lines and appliances is that for power lines, the current is unaffected by the resistance of the line itself, being determined instead by the load or power consumption at the end of the line (this is because the resistance of the line itself is very small and insignificant compared to that of the appliances at the end, so that any reasonable change in the resistance of the line will have a negligible effect on the overall resistance, and thus the current flowing through it).

However, the voltage drop along the line (i.e., the difference in voltage between its end-points, not to be confused with the line voltage relative to ground) is unconstrained and varies depending on current and the line's resistance. Thus, Ohm's law still holds, but it is now I that is fixed and V and R that vary. Applying the formula $P = I^2 \times R$ for resistive heating with the current held constant, we see that doubling the resistance of the power line will double resistive losses.

Since in practice it is desirable to minimize resistive losses on power transmission and distribution lines, these conductors are chosen with the minimal resistance that is practically and economically feasible.

Current Density

Current strength, in a conductor, can be defined as the rate of flow of charge across any cross-section of the conductor. An electric current is usually considered as a flow of electrons. When two ends of a battery are connected to each other by using metal wire, electrons flow out of one end of the battery, through the wire, and into the other end of the battery. Current is usually steady if its magnitude is constant and its direction will always be the same.

Current density refers to the density of current flow in some conductor. It is denoted by the symbol J. In the field of electromagnetism, Current Density and its measurement are very important. It is the measure of the flow of electric charge in amperes per unit area of cross-section i.e. m^2. This is a vector quantity because with the magnitude it is having the direction of flow. An electrical current that flows through and has units of charge per unit time per unit area. It is also measured in the direction which is perpendicular to the flow of direction.

The unit of electric current density is denoted as ampere per square meter. In the study of electromagnetism, Current Density is the measurement of electric current. It concerns the amount of current flowing across the given area.

Suppose a medium (conducting material) has an electric conductivity given by,

$$\sigma$$

Then the electric current density can be related with the Electric Field by the equation,

$$J = \Sigma \times E$$

- J = Current density.
- Sigma = Conductivity.

- E = Electric field.

In another way,

Current Density Formula can be expressed as,

$$j = \frac{I}{A}$$

Where,

J =Current density in A/m^2.

A =The cross-section area in m^2.

I =Current flowing through the conductor in Amperes.

Electric Voltage

Voltage is one of the fundamental parameters associated with any electrical or electronic circuit. Voltage is seen widely in specifications of a host of electrical items from batteries to radios and light bulbs to shavers, and on top of this it is a key parameter that is measured within circuits as well and used within electronic circuit design calculations.

The unit of voltage or potential difference is the volt and it is widely used in all aspects of electrical and electronic circuits and electronic circuit design. Along with current and resistance the unit of voltage is essential in the design and implementation of any circuit.

The operating voltage of an item of equipment is very important - it is necessary to connect electrical and electronic items to supplies of the correct voltage. Connect a 240 volt light bulb to a 12 volt battery and it will not light up, but connect a small 5V USB device to a 240 volt supply and far too much current will flow and it will burn up and be irreparably damaged.

On top of this, the voltage levels within a circuit give a key to its operation - if the incorrect voltage is present, then it may give an indication of the reason for the malfunction. Also many electrical and electronic components have maximum operating voltages so it is very important to stay within their specifications.

For these and many reasons, electrical voltage is a key parameter and knowing what the voltage is can be a key requirement in any circumstance.

Voltage can be considered as the pressure that forces the charged electrons to flow in an electrical circuit. This flow of electrons is the electrical current that flows.

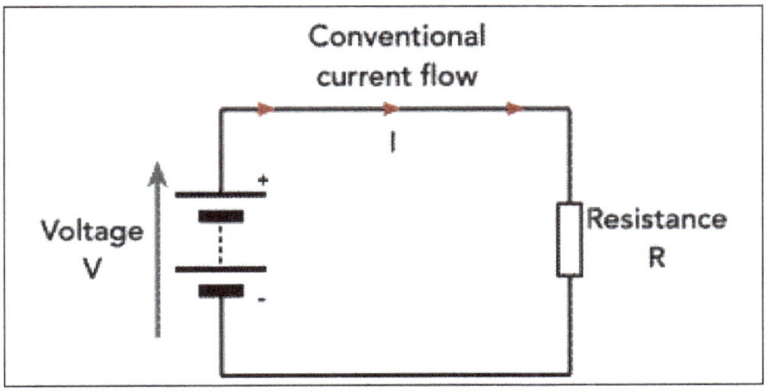

Figure: Voltage shown within a simple circuit.

If a positive potential is placed on one end of a conductor, then this will attract that negative charges to it because unlike charges attract. The higher the potential attracting the charges, the greater the attraction and the greater the current flow.

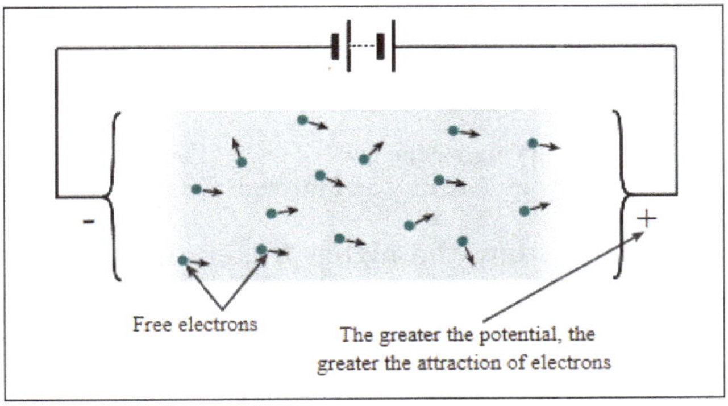

Figure: The higher the voltage potential difference, the greater the attraction of electrons and greater the current flow.

In essence, the voltage is the electrical pressure and it is measured in volts which can be represented by the letter V. Normally the letter V is used for volts in an equation like Ohm's law, but occasionally the letter E may be used - this stands for EMF or electro-motive force.

To gain a view of what voltage is and how it affects electrical and electronic circuits, it is often useful as a basic analogy to think of water in a pipe, possibly even the plumbing system in a house. A water tank is placed up high to provide pressure (voltage) to force the water flow (current) through the pipes. The greater the pressure, the higher the water flow.

Alessandro Volta

The unit of voltage or electrical potential is the volt which is named after Alessandro Volta, an Italian physicist who lived between 1745 and 1827.

Potential Difference

The electrical potential or voltage is a measure of the electrical pressure available to force the current around a circuit. A useful comparison for these purposes is a simple system containing water such as a water tank with a pipe attached and the water passing through a half open tap. The higher the level of water above the tap, the greater the pressure forcing the water through the pipe and through the half open tap. The greater the water pressure, then the more water that will flow through the system for a given level of resistance in the system.

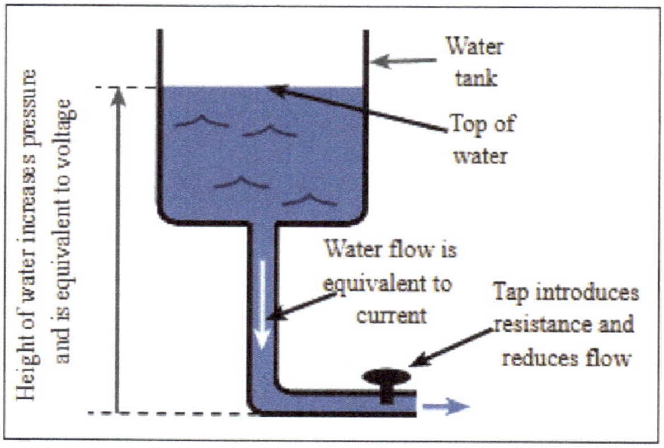

Figure: The higher the level of water, the greater the pressure forcing water through the system.

Similarly with an electrical system, the higher the electric pressure or potential difference across a section of the system, the greater the amount of will pass through the system for a given level of electrical resistance.

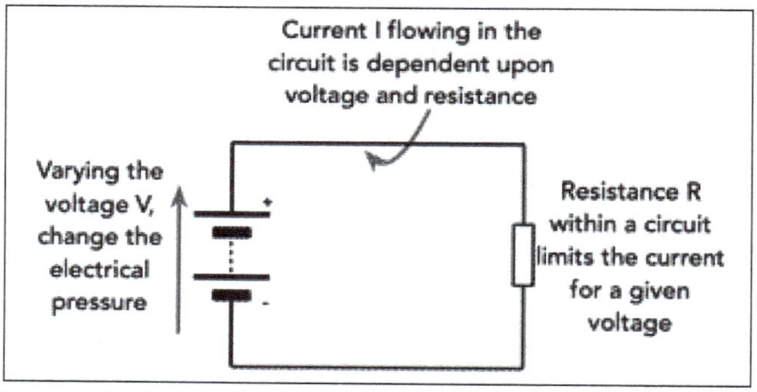

Figure: The higher the electrical pressure or voltage, the higher the current for a given level of resistance.

It can be seen that that raising the pressure of water increases the flow. For an electrical circuit raising the electric potential or voltage increases the current flowing. When looking at the analogy of the water system as an explanation for potential difference, it is worth remembering that it is only a basic analogy, and there are some fundamental differences between the water tank and an electrical circuit, especially in terms of the fact that the electrical circuit is just that, and the water system is not. However it serves well to illustrate the concept of pressure and electric potential in a manner that can be understood.

What is a Volt: Unit of Voltage

The basic unit of voltage is the volt, named after the Italian scientist, Alessandro Volta, who made some early batteries and performed many other experiments with electricity.

The standard unit of voltage or potential difference and electromotive force in the International System of Units (SI), is formally defined to be the difference of electric potential between two points of a conductor carrying a constant current of one ampere, when the power dissipated between these points is equal to one watt.

To give an idea of the voltages which are likely to be encountered, a CB radio will usually operate from a supply of around 12 volts (12 V). The cells used in domestic batteries have a voltage of around 1.5 volts. Rechargeable Nickel Cadmium cells have a slightly smaller voltage of 1.2 volts, but can normally be used interchangeably with the non-rechargeable types.

In other areas voltages much smaller and much greater than this can be encountered. The signal input to an audio amplifier will be smaller than this, and the voltages will often be measured in millivolts (mV) or thousandths of a volt. The signals at the input to a radio are even smaller than this and will often be measured in microvolts (μV) or millionths of a volt.

At the other extreme much greater voltages may be heard about. The cathode ray tubes in a television or computer monitors require voltages of several kilovolts (kV) or thousands of volts, and even larger voltages of millions of volts or megavolts (MV) may be heard of in conjunction with topics like lightning.

EMF and PD

When dealing with voltages, two terms will often be seen: electromotive force, EMF and potential difference, PD. These terms have many similarities, but also some key and very important differences.

To explain this, take the example of a simple battery used to power a circuit. Normally on the outer package of the battery itself the voltage will be stated: often 1.5 volts for single cell alkaline cells and the like. However it is found that when the battery is in use its voltage will fall, especially when larger loads are applied and as it becomes older and it has been used. Even when the cell or battery is new, some voltage drop will be seen when it supplies current. The reason for this is that there is some resistance inside the cell.

From this it is possible to define the EMF as the driving potential within any electrical or electronic source regardless of any internal resistance. It is effectively the voltage seen on the output of the source, i.e. battery, etc. when it is measured with a very high impedance and no load. It is the internal voltage of the cell.

The potential difference at any point is the actual potential see at any given point in a circuit. It is not dependent upon there being no load. In fact the EMF or a source and the potential difference are the same when no load is applied to the source. As a load is applied the potential difference will fall, but the EMF remains the same.

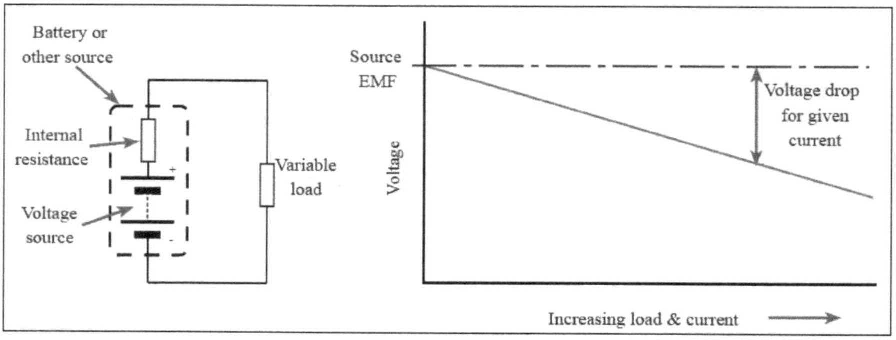

Figure: Voltage and electromotive force, EMF.

How to Measure Voltage

One of the key parameters that need to be known in any electrical or electronic circuit is the voltage. There are several ways in which voltage measurements can be made, but one of the most common is to use a multimeter. Either analogue or digital multimeters can be used, but these days' digital multimeters are most commonly used as they are more accurate and are available for very reasonable prices.

Voltage is one of the three main electrical units along with current and resistance. Voltage is key to the electronic circuit design process as well as that of any electrical circuits. Accordingly it is used in virtually all design processes and a parameter which is associated with very many electrical and electronic components.

Standard Line Voltages

The standard line voltages in use are:

Line Voltages	Usage
240/415V (3 phase)	Used to supply customers installations.
240/480V (1 phase)	
6.6kV	Used for urban and rural HV distribution.
11kV	
22kV	
12.7kV (SWER)	
22kV	
33kV	Used for sub-transmission of larger power levels in distribution over middle distances.
66kV	

110kV	Used for transmission of large power levels over long distance.
220V	
330kV	
500kV	

Voltage between Live Conductors and Voltage to Neutral

The voltage between any two live conductors is often referred to as the "line voltage". The voltage to neutral, often referred to as the "phase voltage", is the voltage between any live conductor and the neutral point or earth of the system.

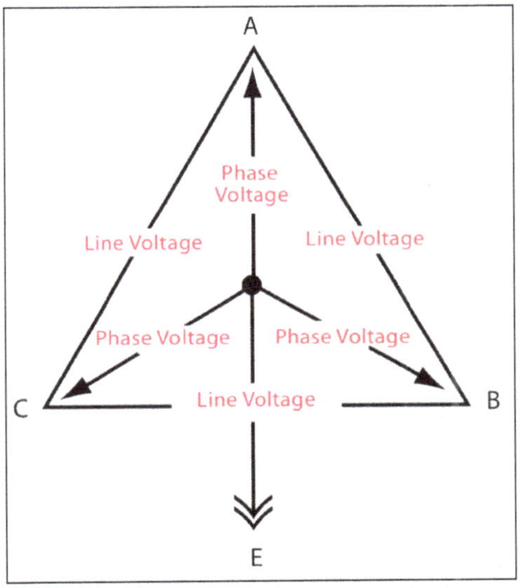

Figure: Three-phase system with earthed neutral.

Figure shows the line and phase voltages in a three-phase system. The neutral point is usually earthed at the supply end (for protection and safety reasons) and each live conductor is then at a definite potential to earth.

For instance, in an 11kV three-phase system, the voltage between any two live conductors gives a line voltage of 11kV while the voltage between any live conductor and neutral (or earth) gives a phase voltage of 6.35kV.

Voltage Systems

High Voltage Overhead Systems

The two systems most commonly used for transmission and distribution are:

- Single-phase.
- Three-phase.

High Voltage Single-phase System

This system is generally associated with the distribution of low power levels over relatively short distances. Single-phase systems are generally fed from a three-phase line. The single-phase line consists of two conductors, neither directly earthed to the general mass of earth. In this system there is no neutral conductor.

It is usual to have the three-phase system earthed (at the neutral point of the transformer or generator supplying the system) either solidly or through some current limiting resistance (for safety and protection purposes). As the single-phase HV system is part of the three-phase HV system, each phase of the single-phase system has a definite voltage to earth.

For safety reasons alone, it is important to remember that each phase is alive to earth and that a definite voltage exists between each phase and the equipment connected to the ground.

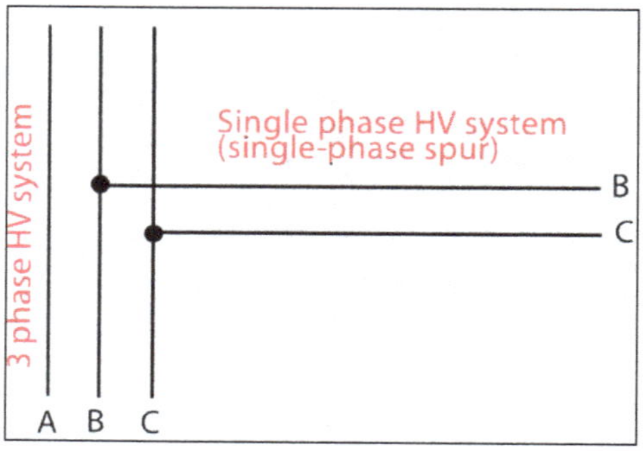

Figure: Three-phase high voltage system with single-phase spur.

High Voltage Three-phase System

This system is widely used for the transmission of high power levels and is also the standard system used in distribution and reticulation.

It consists of three conductors; each called a "phase". To standardize the identification of the phases, they are known as A, B and C phases or red, white and blue phases respectively.

The voltage in each phase alternates, in a similar manner to the alternating voltage shown in Figure but one follows the other in regular order.

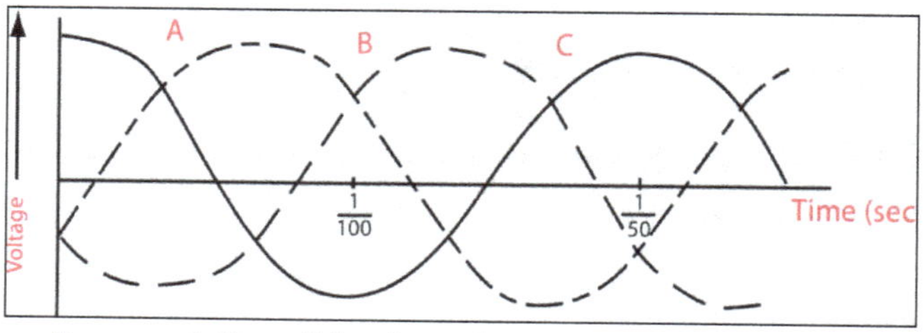

Figure: Representation of the three sine waves in a three-phase system.

Brie y, phase A reaches its maximum positive value first, then is followed by phase B, then by phase C and so on. The order in which the phases reach their peak is called the phase sequence.

Phase Sequence

It is essential that the order of phase sequences and the identity of the A, B and C be known. In the case just cited, the order of phase sequence was from A to B to C because the voltage in phase B reached its maximum value after that in phase A and the voltage in phase C reached its maximum value after that in phase B.

Phase sequence has an important bearing on the direction of rotation of three-phase a.c. motors, which depend on the phase sequence and the relative position of the three-phases connected to the motor terminals.

A reversal in the order of the phase sequence (eg. by interchanging any two of the three wires connected to its main terminals) will cause the motor to run in the reverse direction of rotation.

For this reason alone, it is important that electrical workers know what happens if there is an inadvertent change in the position of the phases supplying a factory in which motors are installed.

Low Voltage Single-phase 2-wire Overhead System

In this system there are two conductors, one generally solidly earthed at the transformer and known as the "neutral", while the other is known as the "live", "active" or "phase" conductor.

The voltage between phase and neutral is nominally 240V and the voltage of the phase or active conductor to earth is therefore also 240V.

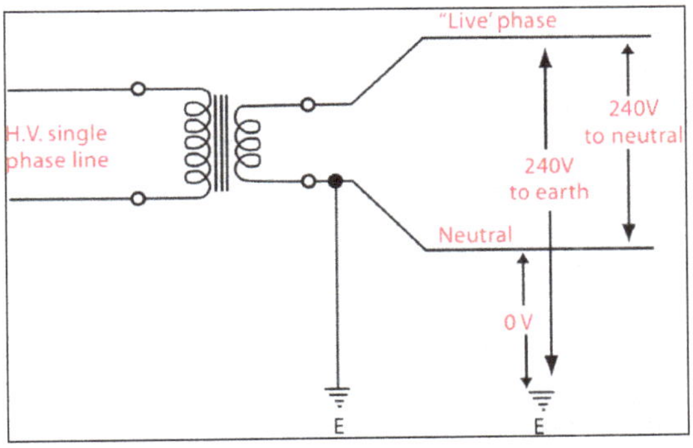

Figure: Single-phase 2-wire system.

Low Voltage Single-phase 3-wire System

In certain rural areas, it is often more economical to install a single-phase high voltage line, saving the cost of the third high voltage phase and to supply the load by stepping down through a transformer to a 3-wire system. One conductor is earthed and known as the neutral while the other conductors are both "actives".

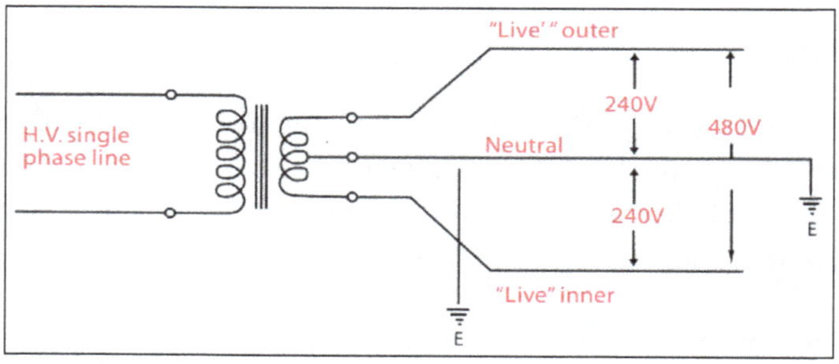

Figure: Single-phase 3-wire system.

The voltage between either of the actives and the neutral is 240V while the voltage between the two active conductors is 480V. It is the a.c. equivalent of the three-wire d.c. system. It facilitates the supply of larger loads or loads at greater distances from the transformer than the single-phase 2-wire system.

Half of a domestic 240V load is connected between one active and the neutral and the other half between the other active and the neutral. This balances the load on each phase and reduces, if not eliminates the residual current in the neutral.

Low Voltage Three-phase 4-wire System

This system employs four conductors and is widely used in all areas where it is considered economical to supply large amounts of energy for industrial and domestic purposes.

The system is shown in Figure – a, b and c are the active conductors and n is the neutral which is connected to the "star point" of the transformer. It is usual for the "star point" to be earthed as shown.

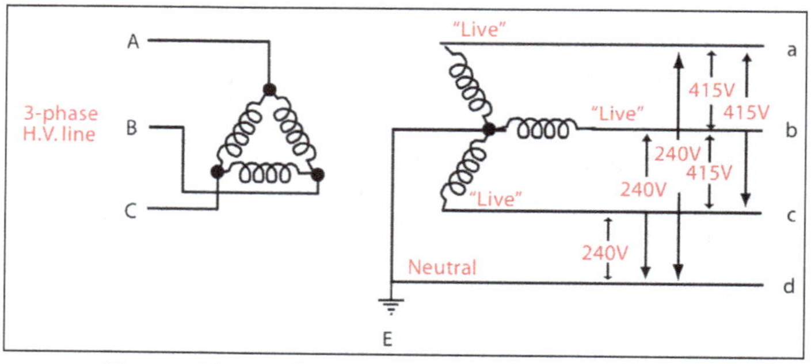

Figure: Three-phase system with earthed neutral.

The standard voltage between actives is 415V, while the voltage between any one of the actives, (a, b and c respectively) and the neutral is 240V.

The same phase relationship of "phase sequence" exists on the LV as on the HV side of the transformer, so care must be taken when renewing mains to avoid upsetting the phase sequence to the supply of motor loads.

High Voltage Single-wire Earth Returns (SWER) System

The power system known as the SWER system uses only one HV conductor with the earth being used as the return conductor. This system was first developed in New Zealand and is now used in Australia, South Africa and many other countries. It can have great economic advantages in hilly areas where the loading is relatively light, where long distances are involved and where the line can be strung from ridge top to ridge top.

Because of the generally lower impedance of the line to earth circuit, it usually has better voltage regulation than a conventional single-phase 2-wire circuit. To restrict noise interference in telecommunications systems, the amount of earth current allowed to own in the earth return circuit is limited. Furthermore, there must be a minimum separation between SWER lines and any telecommunication lines.

A special transformer is used to isolate the SWER line from the main distribution line. The SWER line voltage is 12.7kV to earth. The distribution transformers tied to the SWER line can be either single-phase 2-wire 240V supply or single-phase 3-wire 240/480V supply.

Particular attention must be paid to the good earthing of the transformers on a single-wire line and to the protection of these earth wires from physical damage.

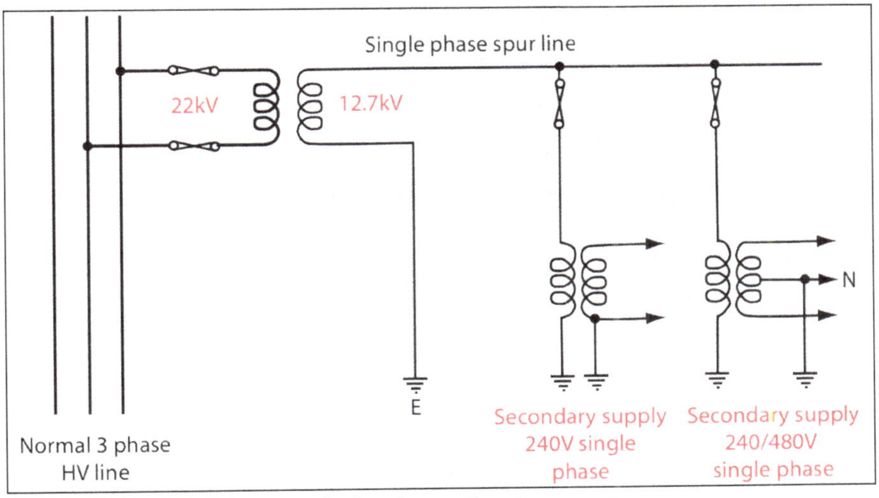

Figure: Single wire earth returns system.

Voltage Optimization

Voltage optimization is the process of systematic reduction in the level of voltages received by a consumer. Ever noticed that all the little gadgets and gizmos you have around the house use slightly different voltages of electricity? Your big appliances are all designed to run off the household supply, typically 110 volts or 230 volts depending on where in the world you are. But the smaller gadgets will use all kinds of different voltages. A flashlight will use about 3 volts, a digital camera 4 volts, a cell phone or CD player 6 volts, a laptop about 20 volts, and so on. Ever wondered why bigger appliances need more? Think voltage, think force: broadly speaking, you need a bigger voltage to force an electric current through something like the electric motor in a refrigerator or a vacuum cleaner than through the tiny little filament lamp in a flashlight or the microchips in your laptop. You need a bigger sledgehammer to crack a bigger nut.

Another way of thinking about this is to remember that the amount of power something electrical uses is proportional to its voltage. If you need an electrical appliance to make your life easy by doing something that requires lots of energy (like trimming your hedge or tumble-drying your jeans), it's also going to need lots of energy putting into it each second—and that means plenty of voltage. In theory, you could power a tumble dryer with a 1.5-volt battery, but its tiny voltage would produce energy much too slowly to evaporate the water from your clothes and such a puny power pack simply doesn't contain enough energy to do the whole job. A 110-volt (or 220-volt) dryer will do the job properly and much faster, but it won't necessarily make any difference if your household power is actually 130 volts (or 250 volts).

Power plants are designed to produce reasonably constant voltages with voltage regulators, ammeters, wattmeters, synchroscopes (which keep power generators in sync with one another and the power grid they're connecting to), and much more.

Overvoltage

Once electricity leaves a power plant, utility companies have little or no idea what we're actually doing with it. They just give us all the same basic supply and let us get on with it. In practice, industrial users will get much higher voltage supplies than homes so they can drive powerful factory machines but, even so, there is only a relatively crude link between the voltage that's supplied and the voltages we actually use. The mismatch between these two things can be a big problem and a huge waste of energy and money.

Most electrical machines are made and sold internationally and have to work on different voltages in different countries. An electric lathe (factory cutting machine) might be manufactured in Germany to work right across Europe (a huge swathe of the world) on voltages ranging from 200–250 volts: in Germany, it'll happily work on 230 volts; in the UK, it'll work just the same (no faster or better) on a national supply that's sometimes closer to 240 volts—but the higher voltage will make it waste about 10 percent more energy by getting considerably hotter (potentially reducing its useful life quite significantly). If you're using that machine in the UK, you might wish your electricity supply were 230 volts instead of 240. This problem is often referred to as overvoltage.

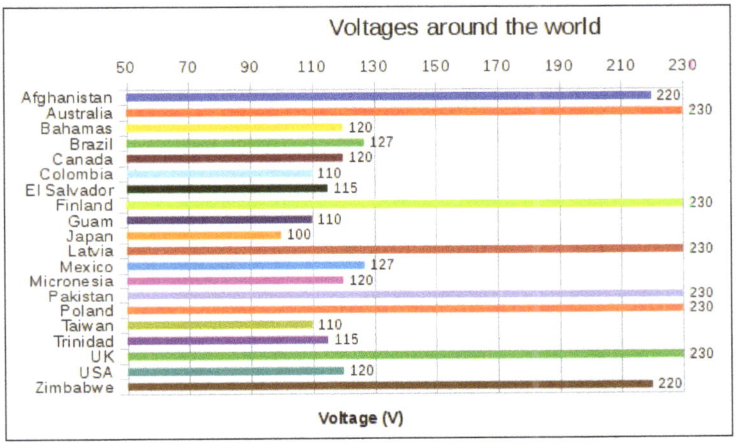

Voltages around the world: As all travellers know, electricity supply voltages vary around the world, though generally there are just two common bands: roughly 100–130 volts for North America and the Pacific and 220–240 volts elsewhere. That's obviously a problem for manufacturers who want to make products for the global market, but less of a problem in a single continent, such as Europe, where voltages have been standardized.

Transients and Harmonics

There are some other issues to worry about too. The voltage your building receives can rise and fall quite dramatically from hour to hour (even from minute to minute or second to second) due to fluctuations in demand and supply. If a factory is turning big electric machines on and off in your neighborhood, for example, that can lead to transients (brief spikes) in power that can affect other buildings nearby. Spikes (sometimes called surges) and sags (sometimes called dips) can also be caused by lightning strikes, power generating equipment going on- or offline, and lots of people all using electricity at once (cooking at the same time each evening, for example). In practice, a supply that is supposed to be 230 volts could be fluctuating regularly by as much as 10 percent or more, giving you an actual voltage anywhere from about 210–250 volts.

And it's not just the voltage that can vary. In theory, most electricity is supplied as an alternating current (AC) sine wave, which rises, falls, and reverses direction smoothly something like 50-60 times a second (the ordinary supply frequency). In practice, AC supplies can also include irregular, higher-frequency waveforms called harmonics that potentially damage delicate equipment (by causing overheating) and really need to be filtered out.

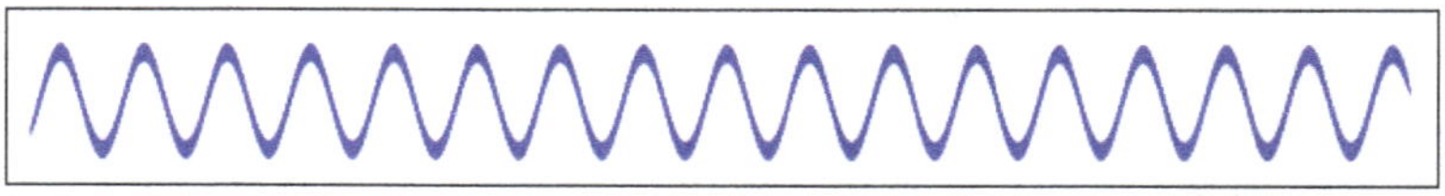

Figure: Ideally, alternating current should vary smoothly in this up-and-down sine-wave pattern. In practice, it can change more drastically and erratically, harming your equipment.

As householders, we don't mind all this too much; most of us aren't even aware of the problem. Like me, you're probably used to having lots of little gadgets (laptops, cell phones, electric toothbrushes, and so on) all with built-in transformers that convert the ordinary power supply (nominally ~110–120 volts in North America and the Pacific region and ~220–240 volts in Europe and elsewhere) to the correct voltage in each case (15 volts for the laptop I'm typing on, for example). And maybe you have surge protectors fitted to a few key appliances (like your computer or your wireless router) to safeguard against spikes and lightning strikes.

If you're running a business, the mismatch between the voltage you're supplied with and the voltage you need is much more of an issue: it's costing you a great deal of money you don't actually need to spend.

Electronic household gadgets that use low voltages typically have small transformers built into their power cords. These all run off the European power supply of 230 volts or so, but actually supply much smaller voltages to the appliances they power. Clockwise from the top: transformers for a modem (18 volts), a cell phone charger (5.9 volts), and an iPod charger (12 volts).

How does Voltage Optimization Work

There are two main ways to solve the problem. The first is to have your own, simple step-down transformer (also called a voltage-reducing or "tap-down" transformer) to change a higher incoming voltage to a lower level more in line with what you actually need. Factories and offices with their own dedicated substation or transformer effectively have this option already; they can simply readjust what's called the "tap setting" (the ratio between the incoming and outgoing voltage) so their transformer supplies a lower voltage than before. Alternatively, an extra step-down transformer can be added to reduce the voltage from the outside electricity supply to one that more closely matches what's needed by the building's internal electricity system. The trouble with this approach is that it solves only the problem of overvoltage. If you're sometimes getting too little voltage from the supply (a problem also called "under voltage" or "brownout"), altering the tap setting to reduce overvoltage will make matters worse.

A step-down transformer like this one (which is supplying my electricity as I type this) reduces the high-voltage power from electricity cables into a lower voltage for homes and factories. Adjusting the tap settings will make it produce a higher or lower output voltage from the same input voltage.

A better solution is to use dedicated voltage optimization equipment that constantly adjusts the voltage from the supply, either increasing or decreasing it so it remains within a narrowly defined band. Devices that do this are called voltage regulators, voltage optimizers, voltage stabilizers, or voltage correctors. They need no maintenance or monitoring and work happily for many years without replacement. They also filter out spikes and harmonics to give a smoother power supply all rounds.

Voltage regulators can work in a variety of different ways. Some are based on ferroresonant transformers (also called constant voltage transformers or CVTs), which are like ordinary transformers but with an extra component (an inductor, based on a capacitor and resonating coil built into their secondary winding). Normally the primary (the "input") and secondary (the "output") of a transformer are coupled so any changes to the primary voltage are directly reflected in the secondary: if the primary voltage rises, the magnetic flux induced in the core of the transformer rises and the secondary voltage rises too by a corresponding amount. The extra circuitry in a ferroresonant transformer keeps the magnetic flux of the secondary section of the transformer at a constant and maximum value, whatever happens to the flux in the primary. That ensures the transformer gives a more or less constant output voltage (usually fluctuating by 1-3 percent) even if the input voltage varies slightly.

Other voltage regulators work a different way. The simplest ones are essentially electronic. They work by constantly measuring the voltage of the waves that make up the incoming electricity supply and comparing them with the voltage you say you want. If there is too much voltage, they add a second wave of just the right size, in antiphase with the original, to subtract exactly the right amount of voltage. So if your incoming supply rises to 250 volts and you'd set the regulator to 220 volts, it will add an upside down waveform equivalent to 30 volts to the 250 volts, subtracting just enough power to make 220 volts. If the supply dips to 240 volts, the corrective voltage will immediately dip to 20 volts, keeping the output steady at 220 volts.

A voltage regulator is usually a more expensive solution than a simple step-down transformer, but it can save half as much energy again, giving overall energy savings of 10-20 percent. Although large units cost thousands of dollars or pounds, typically, they pay for themselves in 2-3 years (in lower energy and maintenance costs and by extending the life of the electrical equipment they're connected to). They also have a payback for the planet: by cutting your energy consumption, they're helping you make a positive contribution to tackling environmental problems such as climate change.

You might think all your household electrical appliances are running off a 110-volt or 220-240-volt supply, but many of them secretly transform that voltage to something else without your knowing. A little circuit like this is built into the base of modern energy-saving fluorescent lamps (also called compact fluorescents or CFLs). It boosts the frequency of the supply so the lamp can be small, bright, and compact and so it doesn't flicker.

Voltage Optimization at Home

Voltage optimization equipment is sophisticated and used to be beyond the budget of all but big businesses and industry. But small, vastly scaled-down versions of the same basic equipment are now being mass-produced for ordinary householders. One small voltage optimization gadget, manufactured by VPhase, promised to reduce household electricity supply from 240-250 volts (potentially fluctuating from 200-250 volts) to a more consistent level around, say, 220 volts, offering a potential saving of 10 percent on electricity bills. Smaller voltage optimization units typically don't protect against transients and harmonics, however and they have been criticized by consumer group which? For costing too much and taking too long to pay for themselves. Ultimately, there are simpler, cheaper, and quicker ways of reducing your household energy bills, some of which (like turning down your thermostats very slightly) don't cost anything at all.

What is the Catch?

Same output, lower input—voltage optimization sounds too good to be true. Indeed, it sounds like it violates one of the most basic laws of physics: the conservation of energy. According to that law, you can't make energy out of thin air or get more energy out of something than you put in. So if you reduce the energy going into an appliance, aren't you going to get less energy out? Or, to put it another way, if voltage optimization lowers the voltage, doesn't it make something like an industrial motor slower and less effective? Won't it make something like a refrigerator or an air conditioner less effective at cooling? Won't it make your lights dimmer?

The answer is sometimes yes, sometimes no. Remember overvoltage, which we introduced above? If you supply something like a motor with more voltage than it needs, it doesn't spin any faster: it just wastes the extra energy as heat. Reduce the voltage and you reduce that wasted heat before you reduce the useful energy that turns the motor. In other words, if you run the motor at its ideal, lower voltage, you make it more efficient. The energy you save in this way is energy you're not drawing from the power supply, so it translates into a financial saving (for you) and an environmental benefit (for the planet). It's worth remembering that if you run appliances with too much voltage or current, they'll wear out significantly more quickly. Extending the lives of electrical appliances also translates into financial and environmental benefits.

Now it's worth pointing out that if you cut the voltage too much, you are bound to reduce the output of whatever appliance you're powering: the conservation of energy tells us that must be the case. Consider an extreme example and this is intuitively obvious: you cannot run a 220-volt domestic refrigerator with a 1.5-volt battery, simply because a battery with that low a voltage can't supply enough energy to power a motor that big. In other words, there is a limit to how much you can cut voltage without cutting useful output.

So what is the limit? Some manufacturers of voltage optimization equipment claim that our everyday voltages are perhaps 10–20 percent too high for the appliances we use, which suggests you can safely cut the voltage you supply by perhaps 10–20 percent without sacrificing any useful output from an appliance. This is misleading. The reality is much less clear-cut and depends on what kind of appliances you're supplying, how they're loaded, and how much you try to reduce the voltage. It's certainly possible to save energy and money with voltage optimization, but you might well reduce the effectiveness of whatever you're powering at the same time.

Will the lights get dimmer with voltage optimization? For many types of lights, there is a roughly linear relationship between the voltage you supply and the light they produce (luminous output); reduce one and you are bound to reduce the other. Think of how a simple flashlight (torch) bulb gradually gets dimmer as the battery loses its charge. What's actually happening (though it's not obvious) is that the battery is slowly losing its voltage (typically from about 1.5 volts, when new, to around 0.5–1.0 volts when it no longer does anything useful) and the lower voltage means less light. Interestingly, for some types of lights, there is also an inverse linear relationship between voltage and lamp life, so the lower the voltage, the longer the light lasts. Now maybe your lights will be a bit dimmer with voltage optimization—and maybe that doesn't matter if your building was too bright to start with or the potential cost savings (through lower power bills and longer-lasting lamps) are more important to you.

Electric Power

The rate at which the work is being done in an electrical circuit is called an electric power. In other words, the electric power is defined as the rate of the transferred of energy. The electric power is produced by the generator and can also be supplied by the electrical batteries. It gives a low entropy form of energy which is carried over long distance and also it is converted into various other forms of energy like motion, heat energy, etc.

The electric power is divided into two types, i.e., the AC power and the DC power. The classification of the electric power depends on the nature of the current. The electric power is sold regarding joule which is the product of the power in kilowatts and the running time of the machinery in hours. The utility of power is measured by the electric meter which records the total energy consumed by the powered devices. The electric power is given by the equation shown below.

$$Electrical\ Power = \frac{Work\ done\ in\ electrical\ current}{time}$$

$$P = \frac{VIt}{t} = VI = IR^2 = \frac{V^2}{R}$$

Where V is the voltage in volts, I is the current in amperes, R is the resistance offered by the powered devices, T is the time in seconds and the P is the power measured in watts.

Unit of Electric Power

The unit of electrical power is Watt.

If,

V=1 Volts and I =1 ampere

P=1 watt

Thus, the power consumed in an electrical circuit is said to one watt if one ampere current flows through the circuit when a potential difference of 1 volt is applied across it.The bigger unit of electrical power is the kilowatt (kW), it is usually used in the power system.

1Kw=1000W

Types of Electric Power

The electrical power is mainly classified into two types. They are the DC power and the AC power.

DC Power

The DC power is defined as the product of the voltage and current. It is produced by the fuel cell, battery and generator.

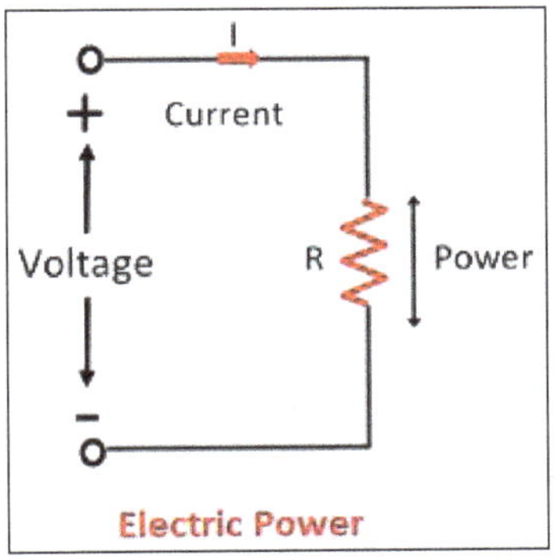

P=V X I

Where P – Power in watt.

V – voltage in volts.

I – current in amps.

AC Power

The AC power is mainly classified into three types. They are the apparent power, active power and real power.

Apparent Power

The apparent power is the useless power or idle power. It is represented by the symbol S, and their SI unit is volt-amp.

$$S = V_{rms} I_{rms}$$

Where S – apparent power.

V_{rms} – RMS voltage = Vpeak√2 in volt.

I_{rms} – RMS current = Ipeak√2 in the amp.

Active Power

The active power (P) is the real power which is dissipated in the circuit resistance.

$$S = V_{rms} I_{rms} \cos \varnothing$$

Where, P – the real power in watts.

V_{rms} – RMS voltage = Vpeak√2 in volts.

I_{rms} – RMS current = Ipeak√2 in the amp.

Φ – impedance phase angle between voltage and current.

Reactive Power

The power developed in the circuit reactance is called reactive power (Q). It is measured in volt-ampere reactive.

$$Q = V_{rms}\, I_{rms}\, Sin\varnothing$$

Where, Q – the reactive power in watts.

V_{rms} – RMS voltage = Vpeak√2 in volt.

I_{rms} – RMS current = Ipeak√2 in the amp.

Φ – impedance phase angle between voltage and current.

The relation between the apparent, active and reactive power is shown below.

$$S^2 = Q^2 + P^2$$

The ratio of the real to the apparent power is called power factor, and their value lies between 0 and 1.

Passive Devices

Passive devices or components do not generate energy, but can store it or dissipate it. Passive devices are the main components used in electronics such as resistors, inductors, capacitors and transformers which together are required to build any electrical or electronic circuit.

As their name suggests, Passive components are electrical components that do not require any form of electrical power to operate, unlike "active devices" such as transistors, operational amplifiers and integrated circuits that require to be powered in some way to make them work.

Being passive, passive devices do not provide gain, amplification or directionality to a circuit but instead provide attenuation as they always have a gain less than one, unity. Therefore passive devices can not generate, oscillate or amplify an electrical signal.

Passive devices can be used individually or connected together within a circuit, either in a series or in a parallel combination to control complex circuits or signals, produce a phase shift to the signal or to provide some form of feedback but they cannot multiply a signal by more than one because they have no power gain. In fact passive devices consume power within an electrical or electronic circuit as they act like attenuators unlike active elements that generate or provide power to a circuit.

The component values of passive devices such as resistance in Ohms or capacitance in Farads are always positive in value (ie, >0) and never negative although some components may have a negative coefficient.

Passive devices are bi-directional components, that is they can be connected either way around within a circuit unless they have a specific polarity marking such as electrolytic capacitors. The polarity of the voltage across them is determined by conventional current flow from the positive to the negative terminal.

Active Devices

An active component is an electronic component which supplies energy to a circuit.

Common examples of active components include:

- Voltage sources.
- Current sources.
- Generators (such as alternators and DC generators).
- All different types of transistors (such as bipolar junction transistors, MOSFETS, FETs, and JFET).
- Diodes (such as Zener diodes, photodiodes, Schottky diodes, and LEDs).

Voltage Sources

A voltage source is an example of an active component in a circuit. When current leaves from the positive terminal of the voltage source, energy is being supplied to the circuit. As per the definition of an active element, a battery can also be considered as an active element, as it continuously delivers energy to the circuit during discharging.

Current Sources

A current source is also considered an active component. The current supplied to the circuit by an ideal current source is independent of circuit voltage. As a current source is controlling the flow of charge in a circuit, it is classified as an active element.

Transistors

Although not as obvious as a current or voltage source – transistors are also an active circuit component. This is because transistors are able to amplify the power of a signal.

Resistive Circuits

Resistive circuits can consist of several resistors in series Power in capacitive circuits can be calculated similar to the way power is calculated in resistive circuits.

Recall that power in resistive circuits is converted to heat and dissipated. That is, the electrical energy is converted into heat energy. This is not true with capacitors. The electrical energy in capacitive circuits is stored temporarily on the plates of the capacitors in what is called an electrostatic field. The energy is then returned to the circuit.

Solving a set of equations that represents a circuit is straightforward, if not always easy. However, developing that set of equations is not so easy. The two commonly taught methods for forming a set of equations are the node voltage (or nodal) method and the loop-current (or mesh) method.

Node Voltage Method

To apply the node voltage method to a circuit with n nodes (with m voltage sources), perform the following steps:

- Selective a reference node (usually ground).
- Name the remaining n-1 nodes and label a current through each passive element and each current source.
- Apply Kirchoff's current law to each node not connected to a voltage source.
- Solve the system of n-1-m unknown voltages.

Example: Consider the circuit shown below.

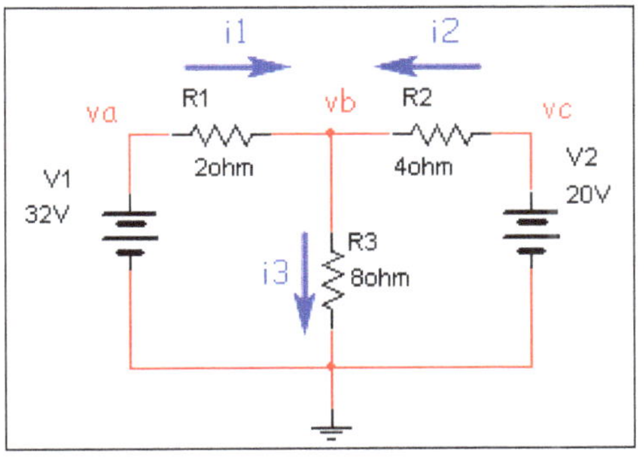

Steps 1 and 2 have already been applied. To apply step 3:

$$i_1 + i_2 = i_3$$

$$i_1 = \frac{v_a - v_b}{R_1},\ i_2 = \frac{v_c - v_b}{R_2},\ i_3 = \frac{v_3}{R_3}$$

$$v_a = V1,\ v_c = V2$$

$$so$$

$$\frac{v_a - v_b}{R_1} + \frac{v_c - v_b}{R_2} = \frac{v_b}{R_3}$$

$$\frac{V1 - v_b}{R_1} + \frac{V2 - v_b}{R_2} = \frac{v_b}{R_3}$$

In this case there is only one unknown, v_b. Plugging in numbers and solving the circuit we get,

$$v_b = 24 Volts$$

The node-voltage method is generally straightforward to apply, but becomes a bit more difficult if one or more of the voltage sources is not grounded.

Example: Consider the circuit shown below.

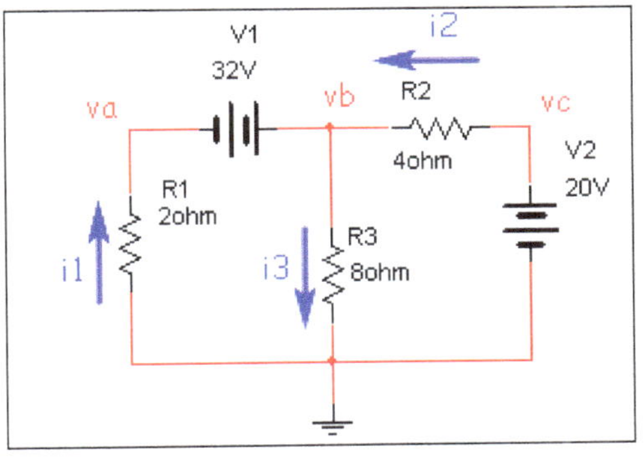

Clearly this circuit is the same as the one shown above, with V1 and R1 interchanged. Now we write the equations:

$$i_1 + i_2 = i_3$$

$$i_1 = \frac{-v_a}{R_1},\ i_2 = \frac{v_c - v_b}{R_2},\ i_3 = \frac{v_3}{R_3}$$

The difficulty arises because the voltage source V1 is no longer identical to one of the node voltages. Instead we have,

$$v_b - v_a = V1,\ v_c = V2$$

$$or$$

$$v_a = v_b - V1$$

$$so$$

$$\frac{-v_a}{R_1} + \frac{v_c - v_b}{R_2} = \frac{v_b}{R_3}$$

$$\frac{V1 - v_b}{R_1} + \frac{V2 - v_b}{R_2} = \frac{v_b}{R_3}$$

Note that the last line is the same as that from the previous circuit, but to solve the circuit we had to first solve for va. This procedure wasn't difficult, but required a little cleverness, and will be a bit different for each circuit layout. Another way to handle this problem is to use the concept of a supernode, which complicates the rules for setting up the equations. However, the supernode concept handles the case of a non-grounded voltage source without any need for solving intermediate equations, as we did here.

The examples chosen here were simple but illustrated the basic techniques of nodal analysis. It also illustrated one of the difficulties with the technique, setting up equations with a floating voltage source. The technique of modified nodal analysis, introduced later, also has no difficulties when presented with floating voltage sources.

Loop Current Method

The loop current (or mesh current) method is, not surprisingly, similar to the node voltage method. To apply the loop current method to a circuit with n loops (and with m current sources), perform the following steps:

- Define each loop current. This is easiest with a consistent method, e.g. all unknown currents are clockwise, all know currents follow direction on current source.

- Apply Kirchoff's voltage law to each loop not containing a current source.
- Solve the system of n-m unknown voltages.

Example: Consider the circuit from Example, with mesh currents defined.

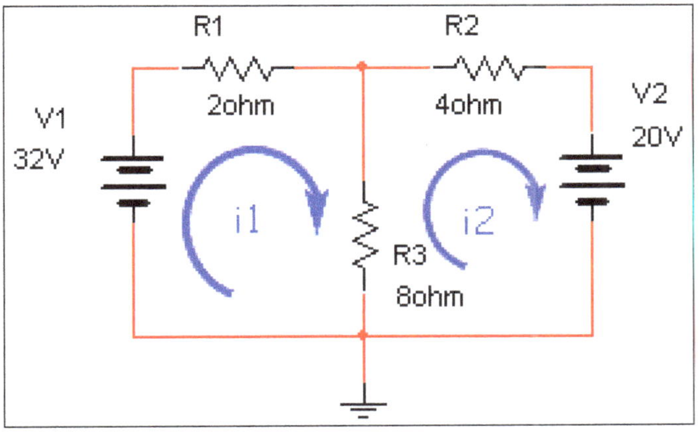

We can apply KVL to both loops,

$$V1 - i_1R_1 - (i_1 + i_2)R_3 = 0$$
$$(i_1 + i_2)R_3 - i_2R_2 - V2 = 0$$

Since there are two equations and two unknowns we can solve by substitution or by matrix methods. To solve by matrix methods we rewrite the equations,

$$(R_1 + R_3)i_1 - R_3i_2 = V1$$
$$R_3i_2 - (R_2 + R_3)i_2 = V2$$

Or,

$$\begin{bmatrix} (R_1+R_3) & -R_3 \\ R_3 & -(R_2+R_3) \end{bmatrix} \begin{bmatrix} i_1 \\ i_2 \end{bmatrix} = \begin{bmatrix} V1 \\ V2 \end{bmatrix}$$

Solving for the two unknown currents we get,

$$i_1 = 4A$$
$$i_2 = 1A$$

While floating current sources tended to complicate the formulation of circuit equations when using the node voltage method, neither the presence of current sources or voltage sources complicates the loop current method.

The choice between the node voltage method and the loop current method is often made on the basis of the circuit at hand. For the example chosen, there was only one independent node but two independent loops. Therefore the node voltage method would be expected to be easier. The situation shown below is the opposite, with two nodes, but only one loop; hence the loop current method is preferable.

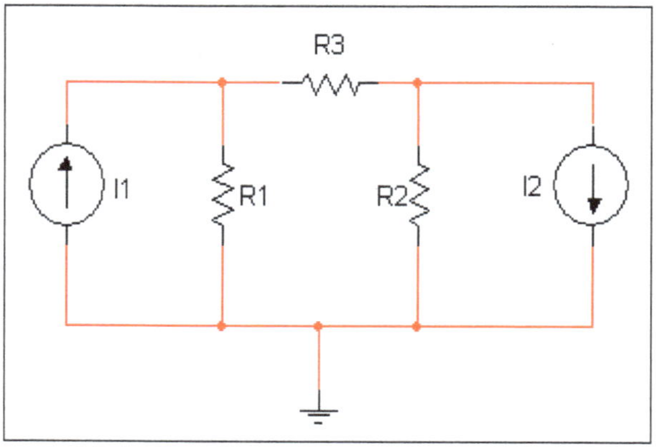

For this circuit you would draw three loops, but two of them go through known current sources - so you would only need one equation. Nodal analysis would require two equations, one each for the voltage on each side of R3.

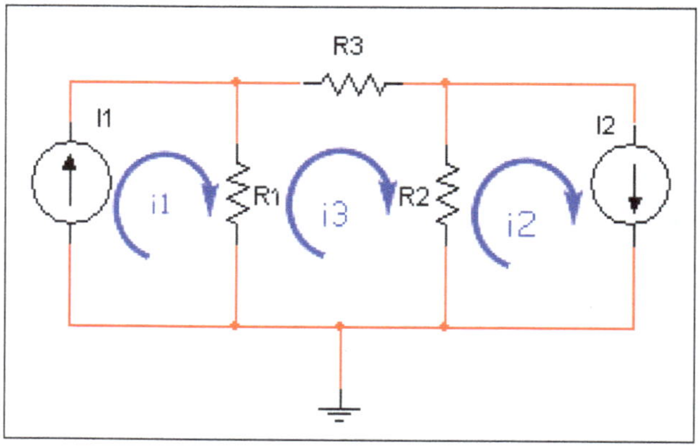

Maximum Power Transfer Theorem

Maximum Power Transfer occurs when the resistive value of the load is equal in value to that of the voltage sources internal resistance allowing maximum power to be supplied. Generally, this source resistance or even impedance if inductors or capacitors are involved is of a fixed value in Ohm´s.

However, when we connect a load resistance, RL across the output terminals of the power source, the impedance of the load will vary from an open-circuit state to a short-circuit state resulting in the power being absorbed by the load becoming dependent on the impedance of the actual power source. Then for the load resistance to absorb the maximum power possible it has to be "Matched" to the impedance of the power source and this forms the basis of Maximum Power Transfer.

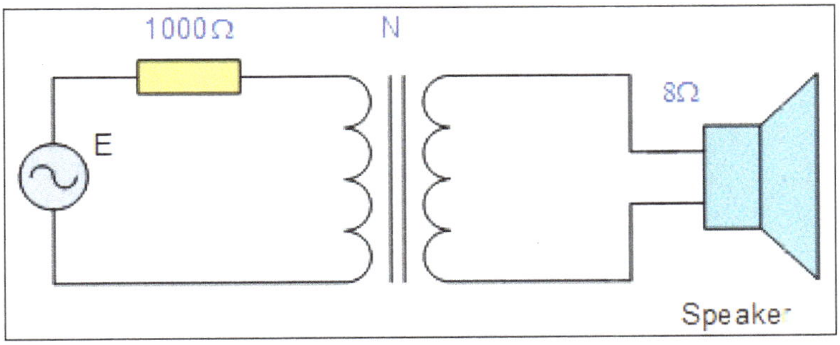

The Maximum Power Transfer Theorem is another useful circuit analysis method to ensure that the maximum amount of power will be dissipated in the load resistance when the value of the load resistance is exactly equal to the resistance of the power source. The relationship between the load impedance and the internal impedance of the energy source will give the power in the load. Consider the circuit below.

Thevenins Equivalent Circuit

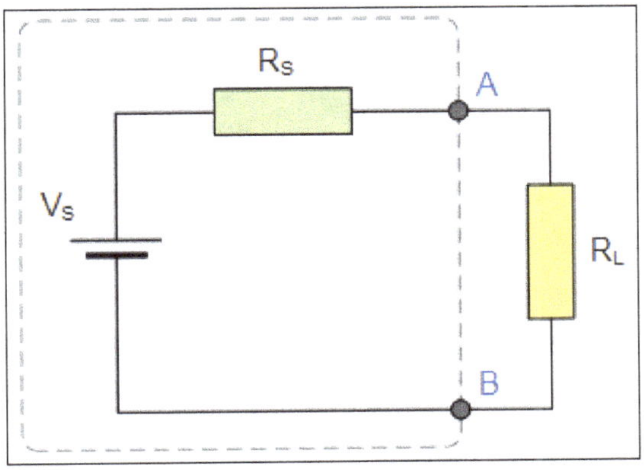

In our Thevenin equivalent circuit above, the maximum power transfer theorem states that "the maximum amount of power will be dissipated in the load resistance if it is equal in value to the Thevenin or Norton source resistance of the network supplying the power".

In other words, the load resistance resulting in greatest power dissipation must be equal in value to the equivalent Thevenin source resistance, then RL = RS but if the load resistance is lower or higher in value than the Thevenin source resistance of the network, its dissipated power will be less than maximum.

For example, find the value of the load resistance, RL that will give the maximum power transfer in the following circuit.

Maximum Power Transfer Example

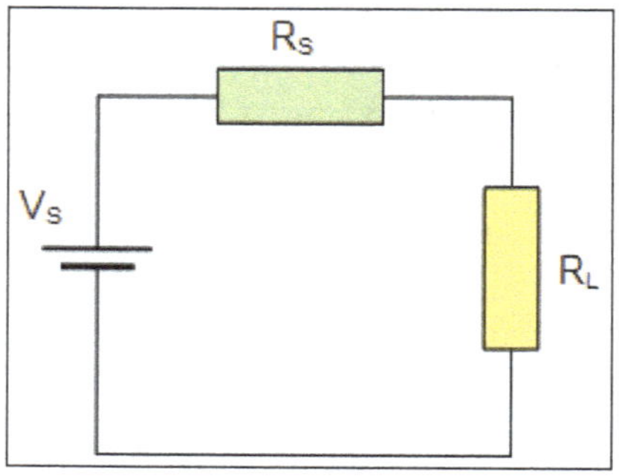

Then by using the following Ohm's Law equations:

$$I = \frac{\quad}{R_S \quad R_L} \quad \text{and } P = I \; R$$

We can now complete the following table to determine the current and power in the circuit for different values of load resistance.

Table of Current against Power.

R_L (Ω)	I (amps)	P (watts)	R_L (Ω)	I (amps)	P (watts)
0	4.0	0	25	2.0	100
5	3.3	55	30	1.8	97
10	2.8	78	40	1.5	94
15	2.5	93	60	1.2	83
20	2.2	97	100	0.8	64

Using the data from the table above, we can plot a graph of load resistance, RL against power, P for different values of load resistance. Also notice that power is zero for an open-circuit (zero current condition) and also for a short-circuit (zero voltage condition).

Graph of Power against Load Resistance

From the above table and graph we can see that the Maximum Power Transfer occurs in the load when the load resistance, RL is equal in value to the source resistance, RS that is: RS = RL = 25Ω. This is called a "matched condition" and as a general rule, maximum power is transferred from an active device such as a power supply or battery to an external device when the impedance of the external device exactly matches the impedance of the source.

One good example of impedance matching is between an audio amplifier and a loudspeaker. The output impedance, Z_{OUT} of the amplifier may be given as between 4Ω and 8Ω, while the nominal input impedance, Z_{IN} of the loudspeaker may be given as 8Ω only.

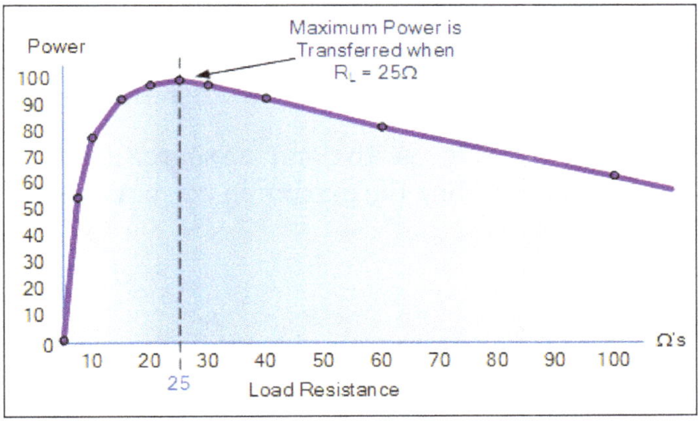

Then if the 8Ω speaker is attached to the amplifiers output, the amplifier will see the speaker as an 8Ω load. Connecting two 8Ω speakers in parallel is equivalent to the amplifier driving one 4Ω speaker and both configurations are within the output specifications of the amplifier.

Improper impedance matching can lead to excessive power loss and heat dissipation. But how could you impedance match an amplifier and loudspeaker which have very different impedances. Well, there are loudspeaker impedance matching transformers available that can change impedances from 4Ω to 8Ω, or to 16Ω's to allow impedance matching of many loudspeakers connected together in various combinations such as in PA (public address) systems.

Transformer Impedance Matching

One very useful application of impedance matching in order to provide maximum power transfer between the source and the load is in the output stages of amplifier circuits. Signal transformers are used to match the loudspeakers higher or lower impedance value to the amplifiers output impedance to obtain maximum sound power output. These audio signal transformers are called "matching transformers" and couple the load to the amplifiers output as shown below.

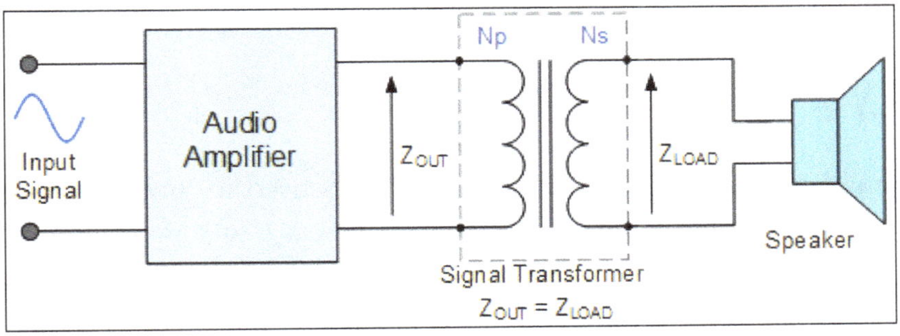

The maximum power transfer can be obtained even if the output impedance is not the same as the load impedance. This can be done using a suitable "turns ratio" on the transformer with the corresponding ratio of load impedance, Z_{LOAD} to output impedance, Z_{OUT} matches that of the ratio of the transformers primary turns to secondary turns as a resistance on one side of the transformer becomes a different value on the other.

If the load impedance, Z_{LOAD} is purely resistive and the source impedance is purely resistive, Z_{OUT} then the equation for finding the maximum power transfer is given as:

$$Z_{out} = \left(\frac{N_P}{N_S}\right)^2 Z_{load}$$

Where: NP is the number of primary turns and NS the number of secondary turns on the transformer. Then by varying the value of the transformers turns ratio the output impedance can be "matched" to the source impedance to achieve maximum power transfer. For example,

Maximum Power Transfer Example

If an 8Ω loudspeaker is to be connected to an amplifier with an output impedance of 1000Ω, calculate the turns ratio of the matching transformer required to provide maximum power transfer of the audio signal. Assume the amplifier source impedance is Z_1; the load impedance is Z_2 with the turns ratio given as N.

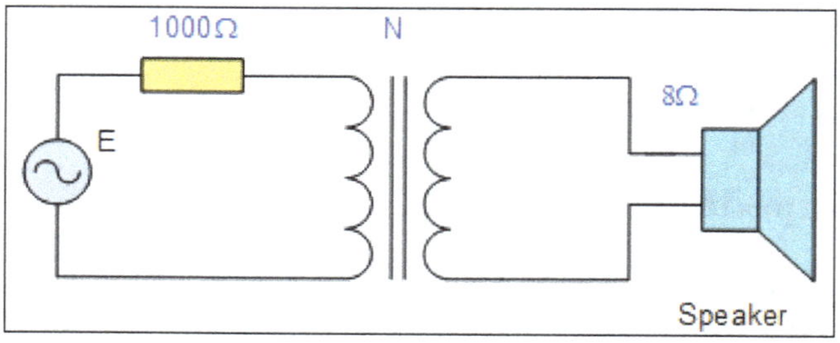

$$Z_1 = N^2 Z_2 \quad \therefore \quad N = \sqrt{\frac{Z_1}{Z_2}}$$

therefore,

$$N = \sqrt{\frac{Z_1}{Z_2}} = \sqrt{\frac{1000}{8}} = 11.2:1$$

Generally, small high frequency audio transformers used in low power amplifier circuits are nearly always regarded as ideal for simplicity, so any losses can be ignored.

References

- What-is-electrical-current, An-electric-current-is-a-protons-positive-ions-or-holes, current, basic-concepts: electronics-notes.com, Retrieved 02, February 2020

- Alternating-current-ac-vs-direct-current-dc: learn.sparkfun.com, Retrieved 10, April 2020
- Resistive-heating-explained-in-details: electrical-engineering-portal.com, Retrieved 07, July 2020
- Current-density-formula, physics-formulas, guides: toppr.com, Retrieved 17, August 2020
- What-is-voltage-basics-tutorial, basic-concepts: electronics-notes.com, Retrieved 15, March 2020
- Voltage-optimisation: explainthatstuff.com, Retrieved 14, January 2020
- Active-amp-passive-components-what-is-the-difference-between-the-two: escomponents.com, Retrieved 18, June 2020

3

Understanding Electrical Circuits

An electrical circuit is a closed loop interconnection of electrical components consisting of electrical elements. The study of electrical circuits is done using circuit diagrams and phasor diagrams. Electrical impedance, network analysis, Ohm's Law, Kirchhoff's Circuit Laws, source transformation, Norton's theorem, compensation theorem, superposition theorem and Thevenin's theorem are a few topics discussed in detail in this chapter.

An electrical circuit is a path or line through which an electrical current flows. The path may be closed (joined at both ends), making it a loop. A closed circuit makes electrical current flow possible. It may also be an open circuit where the electron flow is cut short because the path is broken. An open circuit does not allow electrical current to flow.

Below is a basic set of symbols that you may find on circuit diagrams,

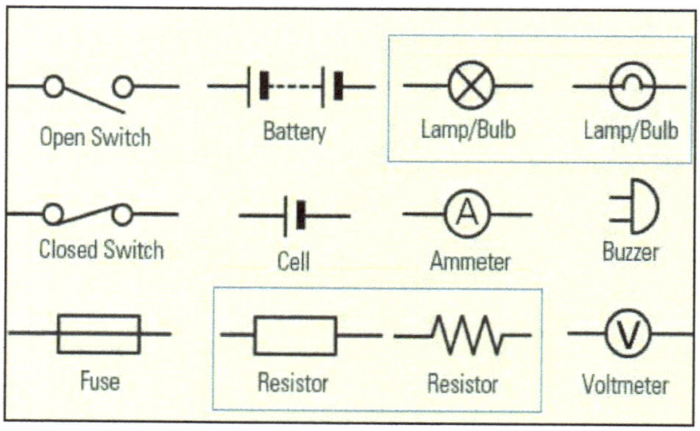

It is very important to know the basic parts of a simple circuit and the symbols that relate to them. A simple circuit has conductors, a switch, a load and a power source. Here are the functions of each part:

Conductors

These are usually copper wires with no insulation. They make the path through which the electricity flows. One piece of the wire connects the current from the power source (cell) to the load. The other piece connects the load back to the power source.

Switch

The switch is simply a small gap in the conductor where you can close or open the circuit. When the switch is closed, the circuit is closed and electricity flows.

The Load

The load is a small light bulb or buzzer that lights when the circuit is turned on. The load is also known as a resistor.

Cell

The power source is a cell. (Note that more than one cell put together is known as a battery).

The diagram below shows how a basic circuit looks like.

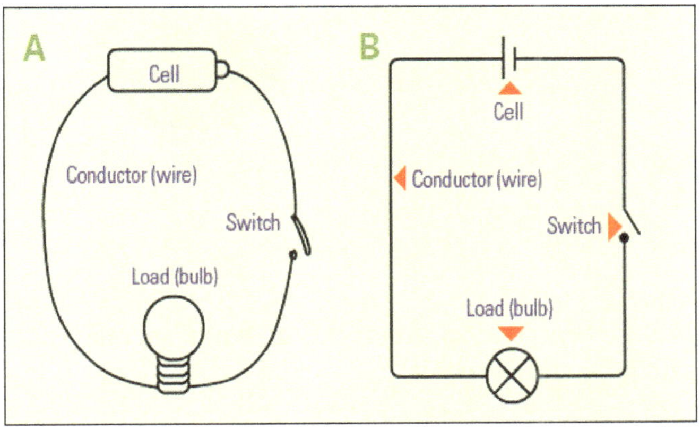

It is important to draw circuits with clean straight lines, as shown in diagram B. Avoid realistic sketches. It is important to know that a circuit can have more than the basic components in the diagram. It can have two or more batteries or two or more bulbs.

Basic Properties of Electric Circuits

- A circuit is always a closed path.
- A circuit always contains at least an energy source which acts as a source of electrons.
- The electric elements include uncontrolled and controlled source of energy, resistors, capacitors, inductors, etc.
- In an electric circuit flow of electrons takes place from negative terminal to positive terminal.
- Direction of flow of conventional current is from positive to negative terminal.
- Flow of current leads to potential drop across the various elements.

Types of Electric Circuit

The electric circuit can be categorized in three different ways:

- Open circuit.
- Closed circuit.
- Short circuit.

Open Circuit

If due to disconnection of any part of an electric circuit if there is no flow of current through the circuit, is said to be an open circuited.

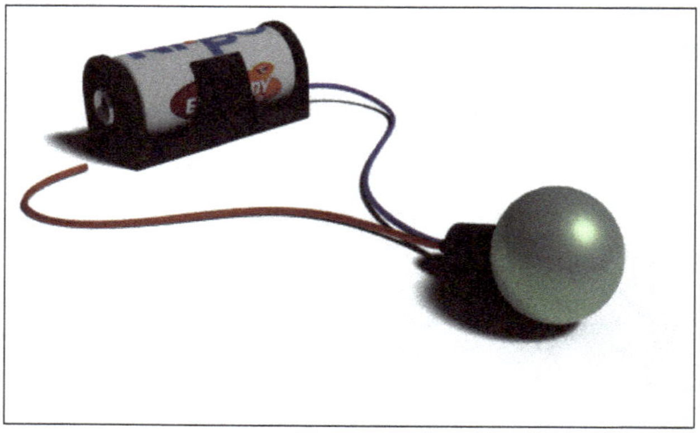

Closed Circuit

If there is no discontinuity in the circuit and current can flow from one part to another part of the circuit, the circuit is said to be closed circuit.

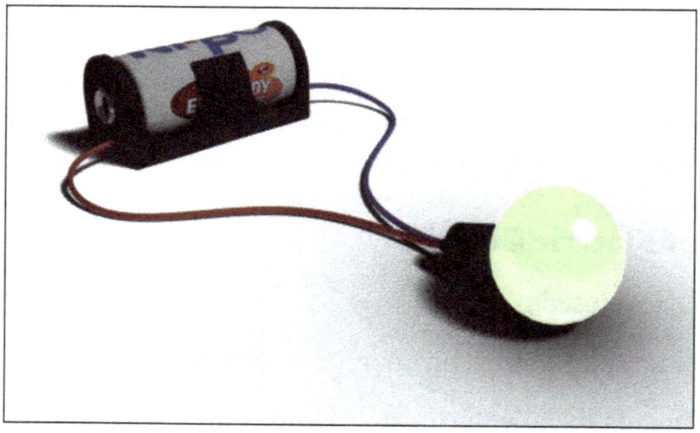

Short Circuit

If two or more phases, one or more phases and earth or neutral of AC system or positive and negative wires or positive or negative wires and earth of DC system touch together directly or connected together by a zero impedance path then the circuit is said to be short circuited.

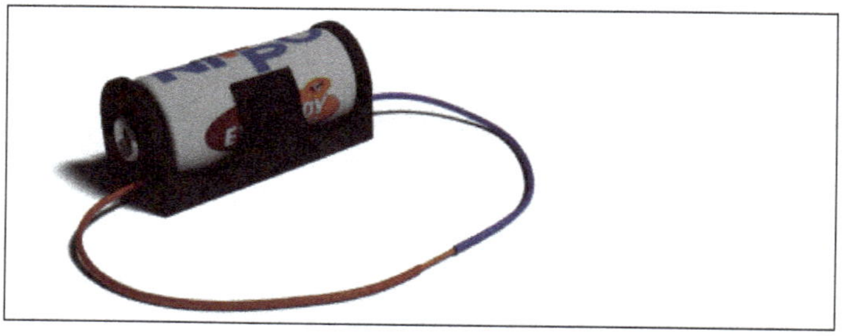

Series Circuit

A series circuit is one that has more than one resistor, but only one path through which the electricity (electrons) flows. From one end of the cell (battery), the electrons move along one path with NO branches, through the resistors, to the other end of the cell. All the components in a series circuit are connected end-to-end.

A resistor in a circuit is anything that uses some of the power from the cell. In the example below, the resistors are the bulbs. In a series circuit, the components are arranged in a line, one after the other.

Take a look at the diagram below:

The schematic drawing is a better way to draw a series circuit.

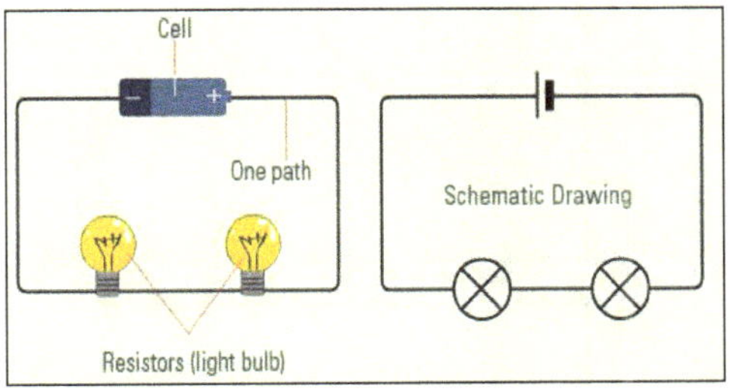

Each time there is damage (break) in any one of the resistors the entire circuit will not function. For example, if one light bulb goes out, all the other lights will go off because the electricity path in the broken bulb is cut off.

Do you put Christmas lights on the trees at home during Christmas? If the lights are in a series circuit, one burned out bulb will keep all the lights off. That is one disadvantage of series circuits. One advantage though is that you will always know if there is a break in a series circuit.

If there are many bulbs in a circuit with a battery (cell), it is very likely that the light will be dimmer because many resistors are acting on the same voltage of power from the battery.

RL Circuit

A circuit that contains a pure resistance R ohms connected in series with a coil having pure inductance of L (Henry) is known as R L Series Circuit. When an AC supply voltage V is applied the current, I flows in the circuit. IR and IL will be the current flowing in the resistor and inductor respectively, but the amount of current flowing through both the elements will be same as they are connected in series with each other. The circuit diagram of RL Series Circuit is shown below:

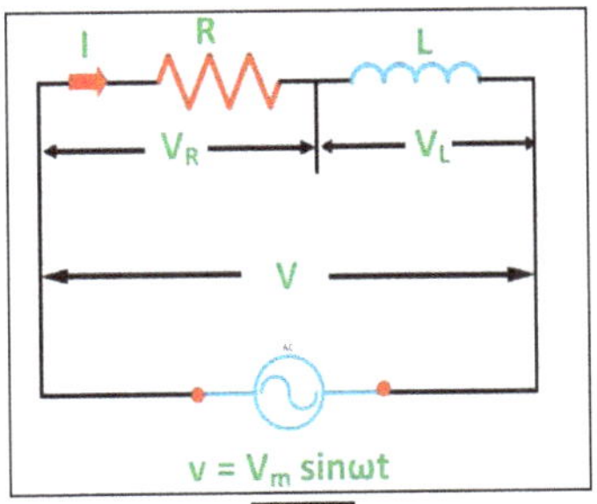

Where,

- V_R – voltage across the resistor R.
- V_L – voltage across the inductor L.
- V – Total voltage of the circuit.

Phasor Diagram of the RL Series Circuit

The phasor diagram of the RL Series circuit is shown below:

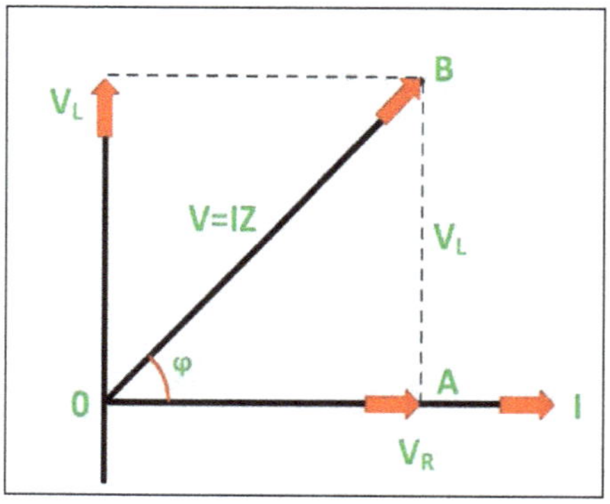

Steps to draw the Phasor Diagram of RL Series Circuit. The following steps are given below which are followed to draw the phasor diagram step by step:

- Current I is taken as a reference.
- The Voltage drop across the resistance $V_R = I_R$ is drawn in phase with the current I.
- The voltage drop across the inductive reactance $V_L = IX_L$ is drawn ahead of the current I. As the current lags voltage by an angle of 90 degrees in the pure Inductive circuit.
- The vector sum of the two voltages drops V_R and V_L is equal to the applied voltage V.

Now,

In right angle triangle OAB:

$$V_R = I_R \textit{ and } V_L = IX_L \textit{ where } X_L = 2\pi fL$$

$$V = \sqrt{(V_R)^2 + (V_L)^2} = \sqrt{(IR)^2 + (IX_L)^2}$$

$$V = \sqrt{R^2\ X_L^2} \quad \text{or}$$

$$I = L = \frac{V}{Z}$$

Where,

$$Z = \sqrt{R^2\ X_L^2}$$

Z is the total opposition offered to the flow of alternating current by an *RL* Series circuit and is called impedance of the circuit. It is measured in ohms (Ω).

Phase Angle

In *RL* Series Circuit the current lags the voltage by 90-degree angle known as phase angle. It is given by the equation,

$$\tan\varphi = \frac{V_L}{V_R} = \frac{IX_L}{IR} = \frac{X_L}{R} \text{ or}$$

$$\varphi = \tan^{-1}\frac{X_L}{R}$$

Power in R L Series Circuit

If the alternating voltage applied across the circuit is given by the equation,

$$v = V_m \text{Sin}\,\omega t$$

The equation of current I is given as,

$$i = I_m \sin(\omega t - \varphi)$$

Than the instantaneous power is given by the equation,

$$p = vi$$

Putting the value of v and i from the $v = V_m \text{Sin}\,\omega t$ and $i = I \sin(\omega t - \varphi)$ equations in the $p = vi$ equation we will get,

$$P = (V_m \text{Sin}\,\omega t) \text{x} I_m \sin(\omega t - \varphi)$$

$$p = \frac{V_m I_m}{2} 2\sin(\omega t - \varphi)\sin\omega t$$

$$P = \frac{V_m}{\sqrt{2}}\frac{I_m}{\sqrt{2}}\left[\cos\varphi - \cos(2\omega t - \varphi)\right]$$

$$P = \frac{V_m}{\sqrt{2}}\frac{I_m}{\sqrt{2}}\cos\varphi - \frac{V_m}{\sqrt{2}}\frac{I_m}{\sqrt{2}}\cos(2\omega t - \varphi)$$

The average power consumed in the circuit over one complete cycle is given by the equation shown below:

$$P = \text{average of } \frac{V_m}{\sqrt{2}}\frac{I_m}{\sqrt{2}}\cos\varphi - \text{average of } \frac{V_m}{\sqrt{2}}\frac{I_m}{\sqrt{2}}\cos(2\omega t - \varphi) \quad or$$

$$P = \frac{V_m}{\sqrt{2}}\frac{I_m}{\sqrt{2}}\cos\varphi - \text{Zero} \qquad or$$

$$P = V_{r.m.s} I_{r.m.s} \cos\varphi = VI\cos\varphi$$

Where, *cos*ϕ is called the power factor of the circuit.

$$\cos\varphi = \frac{V_R}{V} = \frac{IR}{IZ} = \frac{R}{Z}$$

The power factor is defined as the ratio of resistance to the impedance of an AC Circuit.

Putting the value of *V* and *cos*ϕ from the above equation the value of power will be,

$$P = (IZ)(I)(R/Z) = I^2R$$

From the equation above, it can be concluded that the inductor does not consume any power in the circuit.

Waveform and Power Curve of the RL Series Circuit

The waveform and power curve of the RL Series Circuit is shown below:

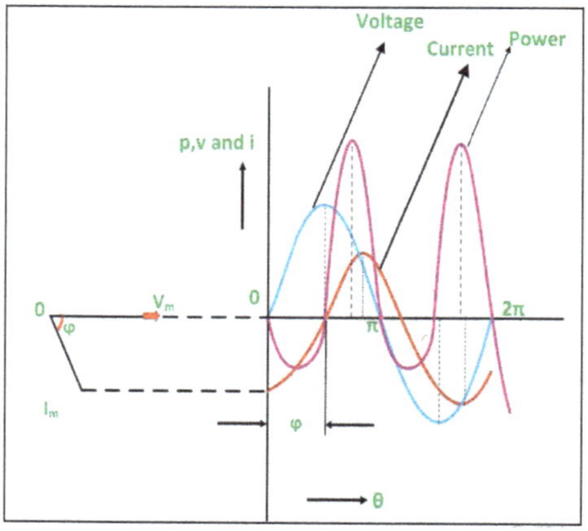

The various points on the power curve are obtained by the product of voltage and current. If you analyze the curve carefully, it is seen that the power is negative between angle 0 and ϕ and between 180 degrees and (180 + ϕ) and during the rest of the cycle the power is positive. The current lags the voltage and thus they are not in phase with each other.

Example:

Calculating Characteristic Time and Current in an RL Circuit

(a) What is the characteristic time constant for a 7.50 mH inductor in series with a 3.00 Ω resistor?

(b) Find the current 5.00 ms after the switch is moved to position 2 to disconnect the battery, if it is initially 10.0 A.

Strategy for (a)

The time constant for an *RL* circuit is defined by $\tau = L/R$.

Solution for (a)

Entering known values into the expression for τ given in $\tau = L/R$ yields.

$$\tau = \frac{L}{R} = \frac{7.50mH}{3.00\Omega} = 2.50ms$$

Discussion for (a)

This is a small but definitely finite time. The coil will be very close to its full current in about ten time constants, or about 25 ms.

Strategy for (b)

We can find the current by using $I = I_0 e^{-t/T}$, or by considering the decline in steps. Since the time is twice the characteristic time, we consider the process in steps.

Solution for (b)

In the first 2.50 ms, the current declines to 0.368 of its initial value, which is,

$$\begin{aligned} I &= 0.368 I_0 = (0.368)(10.0\text{ A}) \\ &= 3.68 A \; at \; t = 2.50 ms \end{aligned}$$

After another 2.50 ms, or a total of 5.00 ms, the current declines to 0.368 of the value just found. That is,

$$\begin{aligned} I' &= 0.368 I = (0.368)(3.68\text{ A}) \\ &= 1.35 A \; at \; t = 5.00 ms \end{aligned}$$

Discussion for (b)

After another 5.00 ms has passed, the current will be 0.183 A; so, although it does die out, the current certainly does not go to zero instantaneously.

RC Circuit

The combination of a pure resistance R in ohms and pure capacitance C in Farads is called RC circuit. The capacitor store energy and the resistor connect in series with the capacitor control the charging and discharging of a capacitor. The RC circuit is used in camera flashes, pacemaker, timing circuit etc.

The RC signal filters the signals by blocking some frequencies and allowing others to pass through it. It is also called first order RC circuit and is used to filter the signals by passing some frequencies and blocking others. The RC fitlers are mostly use for selecting signals and for rejecting noise.

The high-pass filter and the low-pass filter are the most common type of RC filters. The high pass filter passes the frequency greater than the fixed cutoff frequency and blocks the frequency lower than the fixed cutoff frequency. Similarly, the low-pass filter allows the frequency lower than the fixed cutoff frequency and attenuates the frequency higher than the fixed cutoff frequency.

RC Series Circuit

A circuit containing resistance and capacitance connected in series together is called an RC series circuit.

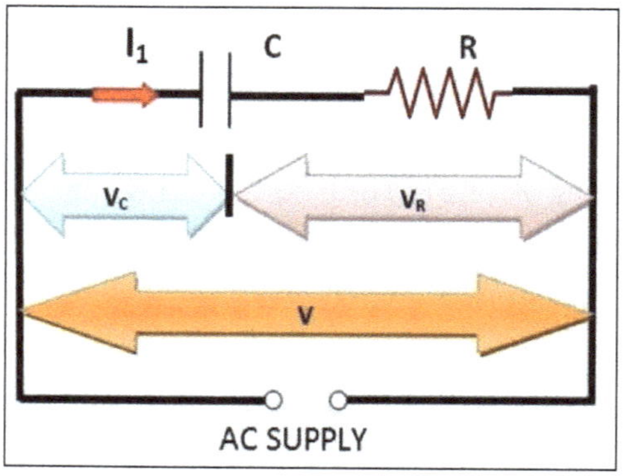

Steps to Draw a Phasor Diagram for an RC Circuit

1. Current I is taken as reference.
2. Voltage drop in resistance is (V_R).

   V_R = IR is drawn in phase with the current I.
3. Voltage drop in capacitive reactance is (V_C):

   V_C = IXC and is drawn 90° behind the current (as current leads the voltage by 90° in pure capacitive load circuit).
4. The vector sum of the two voltage drops is equal to the applied voltage (V).

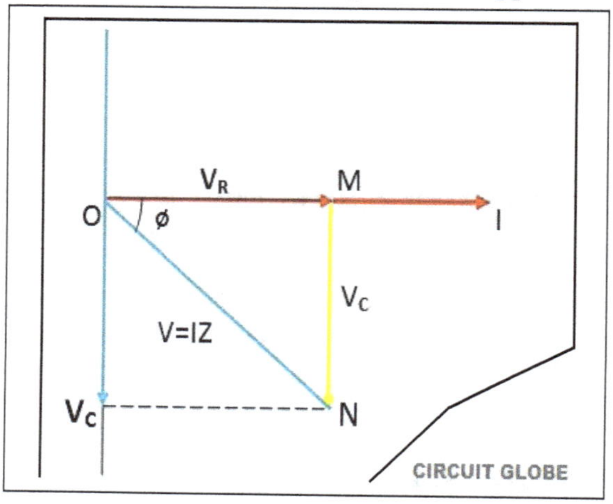

Charging

Figure below shows a simple RC circuit that employs a DC voltage source. The capacitor is initially uncharged. As soon as the switch is closed, current flows to and from the initially uncharged capacitor. As charge increases on the capacitor plates, there is increasing opposition to the flow of charge by the repulsion of like charges on each plate.

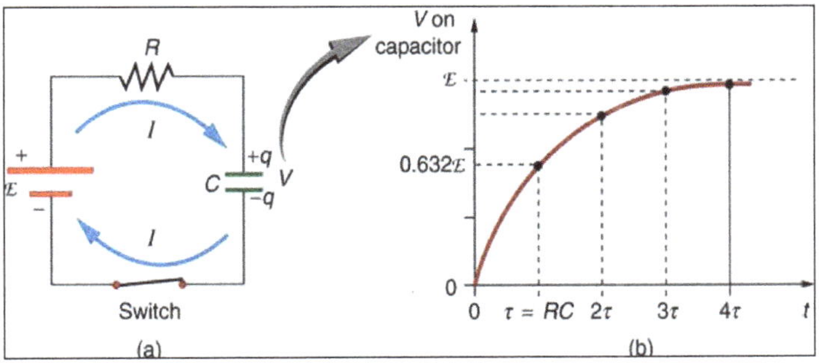

Charging an RC Circuit:

(a) An *RC* circuit with an initially uncharged capacitor. Current flows in the direction shown as soon as the switch is closed. Mutual repulsion of like charges in the capacitor progressively slows the flow as the capacitor is charged, stopping the current when the capacitor is fully charged and $Q=C \cdot emf$.

(b) A graph of voltage across the capacitor versus time, with the switch closing at time $t=0$. (Note that in the two parts of the figure, the capital script *E* stands for emf, *q* stands for the charge stored on the capacitor, and τ is the *RC* time constant).

In terms of voltage, across the capacitor voltage is given by $V_c=Q/C$, where Q is the amount of charge stored on each plate and C is the capacitance. This voltage opposes the battery, growing from zero to the maximum emf when fully charged. Thus, the current decreases from its initial value of $I_0=emf/R$ to zero as the voltage on the capacitor reaches the same value as the emf. When there is no current, there is no IR drop, so the voltage on the capacitor must then equal the emf of the voltage source.

Initially, voltage on the capacitor is zero and rises rapidly at first since the initial current is a maximum. Figure (b) above shows a graph of capacitor voltage versus time (t) starting when the switch is closed at $t=0$. The voltage approaches emf asymptotically since the closer it gets to emf the less current flows. The equation for voltage versus time when charging a capacitor C through a resistor R, is:

$$V(t) = \text{emf}\left(1 - e^{t/RC}\right),$$

where, $V(t)$ is the voltage across the capacitor and emf is equal to the emf of the DC voltage source. (The exact form can be derived by solving a linear differential equation describing the RC circuit, but this is slightly beyond the scope of this Atom.) Note that the unit of RC is second. We define the time constant τ for an RC circuit as τ=RC. τ shows how quickly the circuit charges or discharges.

Discharging

Discharging a capacitor through a resistor proceeds in a similar fashion, as illustrates. Initially, the current is $I_o = V_o/R$, driven by the initial voltage Vo on the capacitor. As the voltage decreases, the current and hence the rate of discharge decreases, implying another exponential formula for *V*. Using calculus, the voltage *V* on a capacitor *C* being discharged through a resistor *R* is found to be,

$$V(t) = V_0 e^{-t/RC}.$$

Impedance

Impedance is the measure of the opposition that a circuit presents to the passage of a current when a voltage is applied.

Rather than solving the differential equation relating to circuits that contain resistors and capacitors, we can imagine all sources in the circuit are complex exponentials having the same frequency. This technique is useful in solving problems in which phase relationship is important. The phase of the complex impedance is the phase shift by which the current is ahead of the voltage.

Complex Analysis

For an RC circuit in, the AC source driving the circuit is given as:

$$V_{in}(t) = Ve^{j\omega t},$$

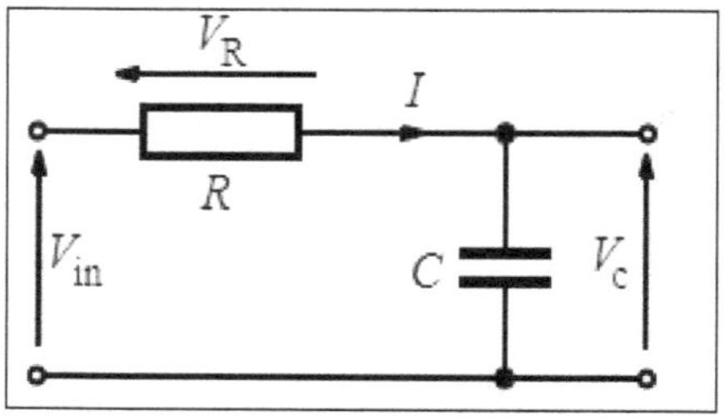

Series RC circuit

where V is the amplitude of the AC voltage, j is the imaginary unit ($j^2=-1$), and ω is the angular frequency of the AC source. Two things to note:

1. We use lower case alphabets for voltages and sources to represent that they are alternating (i.e., we use $v_{in}(t)$ instead of $V_{in}(t)$).
2. The imaginary unit is given the symbol "j", not the usual "i". "i" is reserved for alternating currents.

Complex Impedance

The advantage of assuming sources take this form is that all voltages and currents in the circuit are also complex exponentials (having the same frequency as the source). To appreciate the reason for this, we can investigate how each circuit element behaves when either the voltage or current is a complex exponential. For the resistor, $v=Ri$. From our voltage given above, $i = \frac{V}{R} e^{j\omega t}$. Thus the resistor's voltage is a complex, as is the current with an amplitude $i = \frac{V}{R}$. For a capacitor, $i = C\frac{dv}{dt}$. Letting the voltage be a complex exponential we have $i = j\omega CVe^{j\omega t}$. The amplitude of this complex exponential is $I = j\omega CV$.

The major consequence of assuming complex exponential voltage and currents is that the ratio $Z = \frac{V}{I}$ for rather than depending on time each element depends on source frequency. This quantity is known as the element's (complex) impedance. The magnitude of the complex impedance is the ratio of the voltage amplitude to the current amplitude. Just like resistance in DC cases, impedance is the measure of the opposition that a circuit presents to the passage of a current when a voltage is applied. The impedance of a resistor is R, while that of a capacitor (C) is $\frac{1}{j\omega C}$. In the case of the circuit in, to find the complex impedance of the RC circuit, we add the impedance of the two components, just as with two resistors in series: $Z = R + \frac{1}{j\omega C}$.

Finding Real Currents and Voltages

Since $e^{j\omega t} = \cos(\omega t) + j\sin(\omega t)$, to find the real currents and voltages we simply need to take the real part of the *i(t)* and *v(t)*. The (real value) impedance is the real part of the complex impedance Z. For a series RC circuit, we get $Z = \sqrt{R^2 + \left(\frac{1}{\omega C}\right)^2}$. We see that the amplitude of the current will be $V/Z = \sqrt{R^2 + \left(\frac{1}{\omega C}\right)^2}$.

Phase Angle and Power Factor

In a series RC circuit connected to an AC voltage source, voltage and current maintain a phase difference.

Impedance is an AC (alternating current) analogue to resistance in a DC circuit. As we studied in a previously Atom ("Impedance"), current, voltage and impedance in an RC circuit are related by an AC version of Ohm 's law: $I = \frac{V}{Z}$, where I and V are peak current and peak voltage respectively, and Z is the impedance of the circuit.

In a series RC circuit connected to an AC voltage source as shown in figure, conservation of charge requires current be the same in each part of the circuit at all times. Therefore we can say: the currents in the resistor and capacitor are equal and in phase. (We will represent instantaneous current as *i(t)*.)

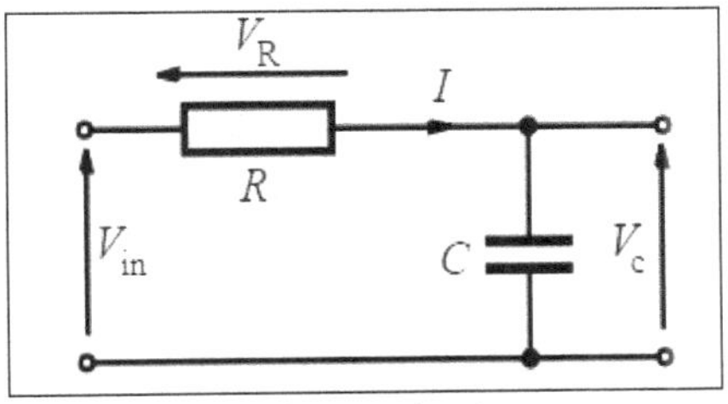

Series RC circuit

On the other hand, because the total voltage should be equal to the sum of voltages on the resistor and capacitor, so we have:

$$\begin{aligned} v(t) &= v_R(t) + v_C(t) \\ &= i(t)R + i(t)\frac{1}{j\omega C}, \\ &= i(t)\left(R + \frac{1}{j\omega C}\right) \end{aligned}$$

where, ω is the angular frequency of the AC voltage source and j is the imaginary unit; $j^2=-1$.

Since the complex number $Z = R + \frac{1}{j\omega C} = \sqrt{R^2 + \left(\frac{1}{\omega C}\right)^2}\, e^{j\phi}$ has a phase angle ϕ that satisfies $\cos\phi = \frac{R}{\sqrt{R^2 + \left(\frac{1}{\omega C}\right)^2}}$,

we notice that voltage v(t) and current i(t) has a phase difference of ϕ.

For R=0, ϕ=90°. As learned from the preceding series of Atoms—the voltage across the capacitor V_C follows the current by one-fourth of a cycle (or 90°).

Power Factor

Because voltage and current are out of phase, power dissipated by the circuit is not equal to: (*peak voltage*) times (*peak current*). The fact that source voltage and current are out of phase affects the power delivered to the circuit. It can be shown that the average power is $I_{rms}V_{rms}\cos\phi$, where I_{rms} and V_{rms} are the root mean square (rms) averages of the current and voltage, respectively. For this reason, $\cos\phi$ is called the power factor, which can range from 0 to 1.

Parallel Circuit

In a parallel circuit, there is more than one resistor (bulb) and they are arranged on many paths. This means electricity (electrons) can travel from one end of the cell through many branches to the other end of the cell.

Look at the figure below involving two resistors in a parallel circuit:

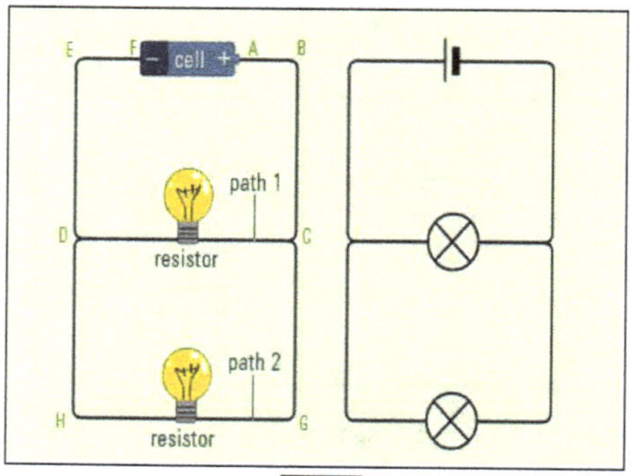

You will notice from the above that there is more than one path:

PATH 1: A-B-C-D-E-F and back to A.

PATH 2: A-B-C-G-H-D-E-F and back to A.

From the above, it is clear that electricity from the cell can take either path A or Path B to return to the cell. The great thing about parallel circuits is that, even when one resistor (bulb) burns out, the other bulbs will work because the electricity is not flowing through one path.

Think of all the light bulbs in your home. If one bulb burns out, the other bulbs in the rooms still work.

Another great thing is that the bulbs in a parallel circuit do not dim out like the case in series circuits. This is because the voltage across one branch is the same as the voltage across all other branches.

Circuit Elements

Circuit elements are those components of an electric circuit or electronic circuit which either can supply energy to the circuit or can take energy from the circuit. All the components of an circuit are connected together in required manner by conductors to form a desired and complete circuit. If the connecting conductor are taken as ideal, all the parts of the circuit, except the connecting conductors are circuit elements.

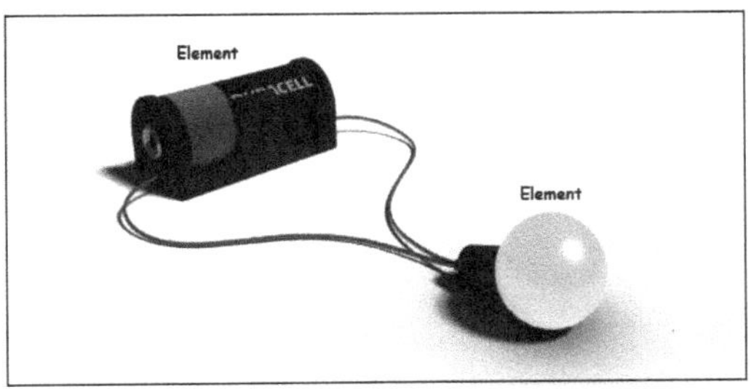

Types of Electric Circuit Elements

The electrical circuit elements either absorb energy from circuit or deliver energy to the circuit. Based on that there are two types of electric circuit elements, namely passive circuit elements and active circuit elements.

Passive Element

The circuit elements that receive energy (or absorb energy) and either convert it into heat or store it in an electric field or a magnetic field are called passive circuit elements.

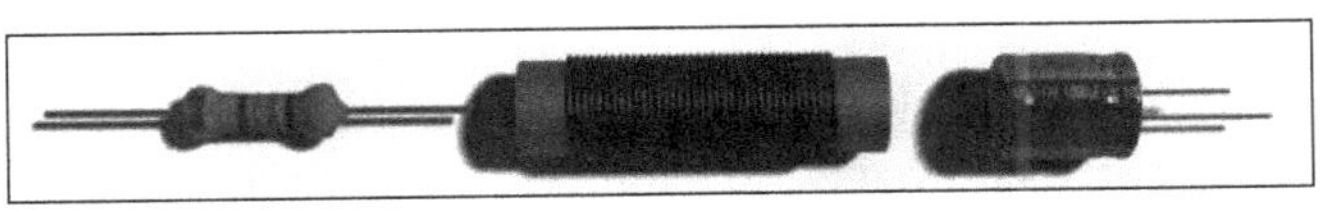

Resistor

Resistor is taken as passive element since it can dissipate energy as heat as long as current flows through it but in any situation a resistor cannot deliver energy to the circuit.

Inductor

An inductor is also considered as passive element of circuit, because it can store energy in it as magnetic field, and can deliver it to the circuit, but not in continuous basis. The energy absorbing and delivering capacity of an inductor are limited and transient in nature. That is why, an inductor is taken as passive element of a circuit.

Capacitor

For same reason, a capacitor is considered as passive element, because it can store energy in it as electric field and deliver it to the circuit, but not in continuous basis. The energy dealing capacity of a capacitor is limited and transient too.

Active Circuit Element

The circuit elements which supply energy to the circuit are called active circuit element.

Examples: Include voltage sources, current sources, and generators such as alternators, DC generators etc.

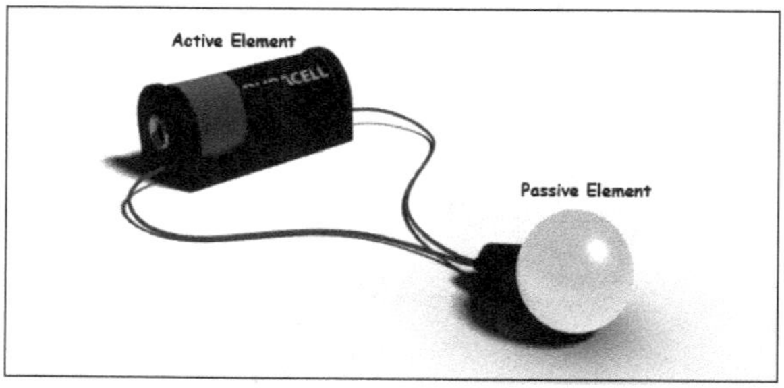

Voltage Source

A voltage source can be considered as an active element. If current leaves from the positive terminal of the voltage source, the energy is delivered from the source to the circuit.

As per definition of active element of electrical circuit, a battery can be considered as active element, as it continuously delivers energy to the circuit during discharging.

Current Source

A current source is also considered as active element. The current supplied to the circuit by an ideal current source is independent of circuit voltage.

Note: A transistor is an active circuit element, because, it can amplify the power of a signal. On the other hand,transformer is not an active circuit element because it does not amplify power and the power remains same both in primary and secondary side. Although, a transistor is an example of active circuit elements, but the transformer is an example of passive circuit elements.

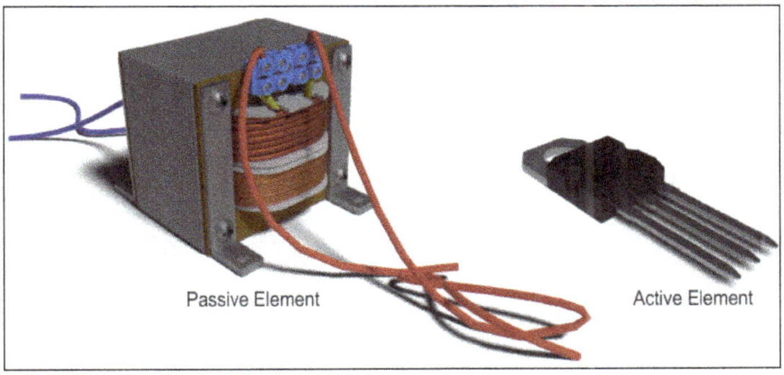

The circuit elements can also be categorised as bilateral and unilateral.

Bilateral Circuit Element

Conduction of current in both directions in a circuit element with same magnitude is termed as bilateral circuit element. It offers same resistance to the current of either directions.

Example: resistor, inductor, capacitor etc.

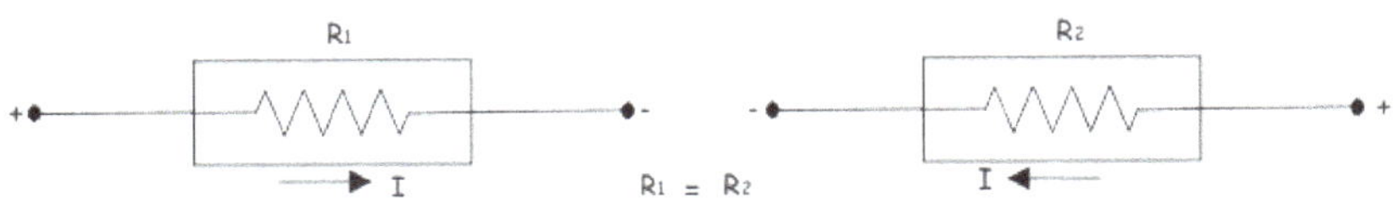

This figure shows a bilateral circuit element can conduct from both sides and offers same resistance for current from either side.

Unilateral Circuit Element

Conduction of current in one direction is termed as unilateral circuit element. It does not offer same resistance to the current of either direction. The resistance of the unilateral circuit element is different for forward current than that of reverse current.

Example: diode, transistor etc.

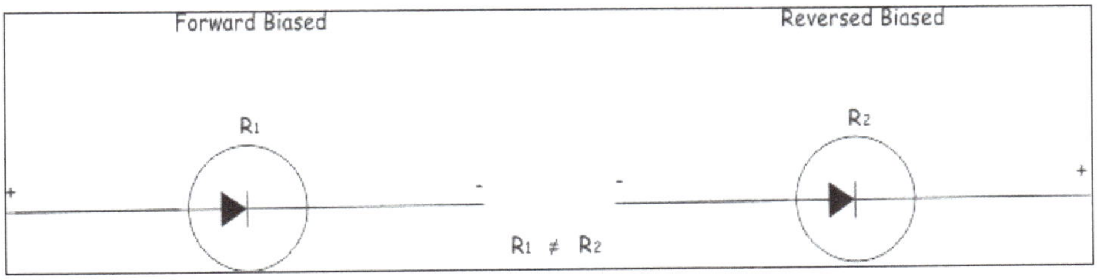

This figure shows, a diode as unilateral circuit element. When diode is forward biased it offers very small resistance and conducts. While it is reverse biased, it offers very high resistance and doesn't conduct. The circuit element can be categorized in another manner, such as lumped and distributed circuit elements.

Lumped Circuit Elements

When the voltage across and current through the element don't vary with dimension of the element, it is called lumped circuit elements.

Example: Resistor connected in any electrical circuit.

Distributed Circuit Element

When the voltage across and current through the element change with dimensions of the element, it is called distributed circuit element.

Example: Resistance of the transmission line. It varies with the length of the line.

Circuit Diagram

Circuit diagram is a simplified conventional graphical representation of an electrical circuit, showing how electrical components are connected together. It's usually used by engineers and electricians to explain symbolically parts and paths of an electrical circuit, which is important in design, construction, and maintenance of electrical and electronic equipment.

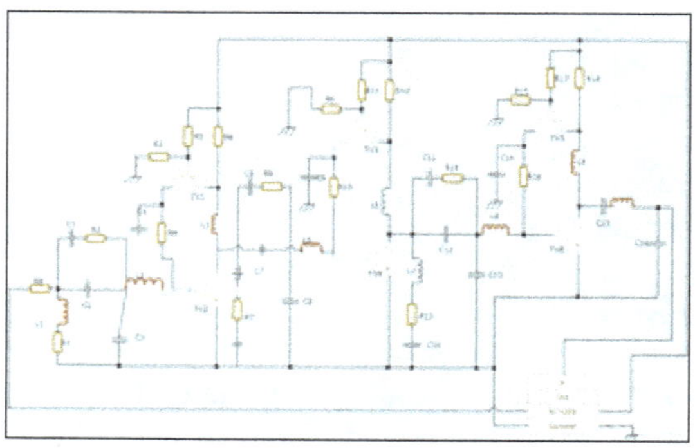

Standard Circuit Symbols for Designing

Circuit diagrams use international standard symbols to provide schematic diagrams of the circuit and its components. Each symbol is intended to represent some feature of the physical construction of the device. First you need to identify the symbols used to represent the components you wish to draw. Edraw provides you with lots of circuits symbols including cell, battery, source, resister, switch, lamp, motor, bell, buzzer, relay, capacitor, etc.

Below are some frequently used circuits symbols:

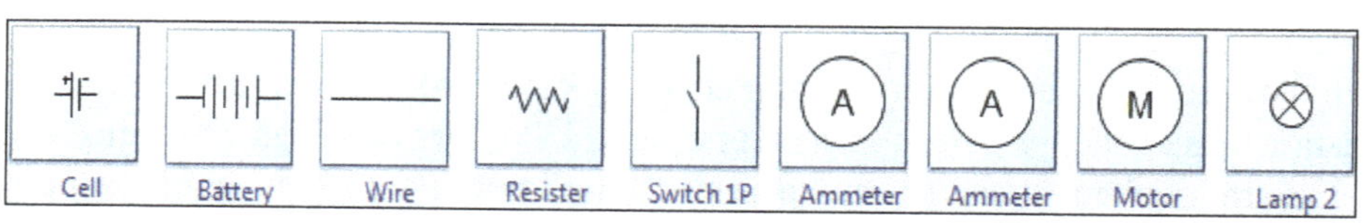

- Cell: A cell stands for electrical energy supply. A single cell is often called a battery, but strictly a battery is two or more cells joined together. It's represented by a long and a short parallel line.
- Battery: A battery is more than one cell. The larger terminal (on the left) is positive (+).It is represented by a collection of long and short parallel lines.
- Wire: Pass current from one part of the circuit to another. It's a connecting wire between two components.
- Resister: A resister is used to restrict the flow of a current. It's represented by a zigzag line.
- Switch: An on-off switch allows current to flow only when it is in the closed (on) position. It's generally represented by providing a break in a straight line by lifting a portion of the line upward at a diagonal.
- Ammeter: An ammeter is used to measure current. It's represented by a letter A in a circle.
- Voltmeter: A voltmeter is used to measure voltage. It's represented by a letter V in a circle.
- Motor: A transducer converts electrical energy to kinetic energy. It's represented by a letter M in a circle.
- Lamp: A transducer which converts electrical energy to light.

Types of Circuit Diagram

Circuit diagrams can be divided into two categories - pictorial circuit diagram and schematic circuit diagram.

Pictorial Circuit Diagram

Pictorial diagrams are much easier to understand than schematic circuit diagrams. Connecting realistic electrical components with the wiring, a pictorial diagram makes it easy for viewers to identify the electrical components of a system immediately without professional knowledge required.

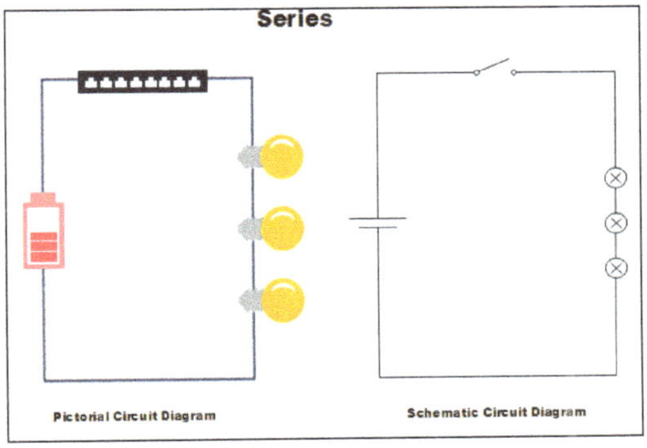

Schematic Circuit Diagram

A schematic circuit diagram represents the electrical system in the form of a diagram that shows the main features or relationships but not the details. In a schematic circuit diagram, the presentation of electrical components and wiring does not completely correspond to the physical arrangements in the real device. If you want to understand a schematic diagram, you are required to master basic knowledge of electricity and physics as well as internationally standardized symbols. Look at the parallel circuits below, you may find that the battery is represented as two short lines, the lights are represented as a circle with a cross inside, and the wiring is represented as a line. This kind of circuit diagram is mainly used by electrical engineers. The following is a schematic circuit diagram of a semiconductor electronic. Schematics and circuit diagrams are both important engineering diagrams.

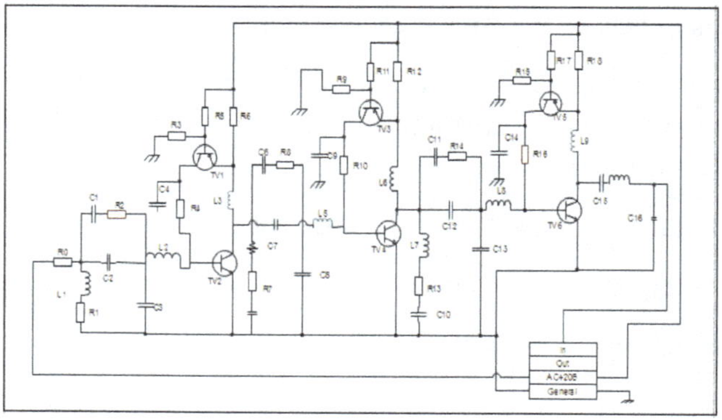

Phasor Diagrams

Phasor diagrams are a graphical way of representing the magnitude and directional relationship between two or more alternating quantities sinusoidal waveforms of the same frequency can have a Phase Difference between themselves which represents the angular difference of the two sinusoidal waveforms. Also the terms "lead" and "lag" as well as "in-phase" and "out-of-phase" are commonly used to indicate the relationship of one waveform to the other with the generalized sinusoidal expression given as: $A_{(t)} = A_m \sin(\omega t \pm \Phi)$ representing the sinusoid in the time-domain form.

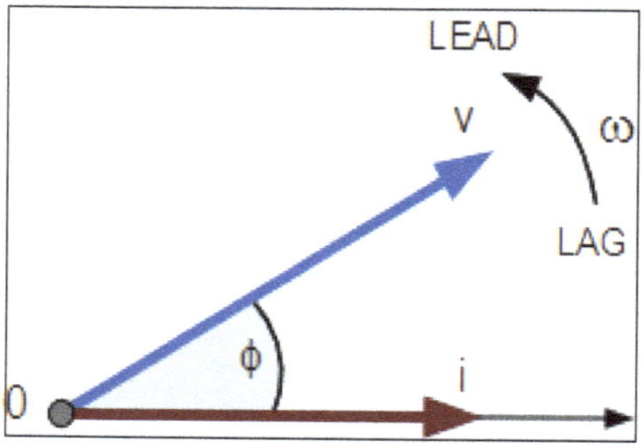

But when presented mathematically in this way it is sometimes difficult to visualise this angular or phasor difference between two or more sinusoidal waveforms. One way to overcome this problem is to represent the sinusoids graphically within the spacial or phasor-domain form by using Phasor diagrams, and this is achieved by the rotating vector method.

Basically a rotating vector, simply called a "Phasor" is a scaled line whose length represents an AC quantity that has both magnitude ("peak amplitude") and direction ("phase") which is "frozen" at some point in time.

A phasor is a vector that has an arrow head at one end which signifies partly the maximum value of the vector quantity (V or I) and partly the end of the vector that rotates.

Generally, vectors are assumed to pivot at one end around a fixed zero point known as the "point of origin" while the arrowed end representing the quantity, freely rotates in an anti-clockwise direction at an angular velocity, (ω) of one full revolution for every cycle. This anti-clockwise rotation of the vector is considered to be a positive rotation. Likewise, a clockwise rotation is considered to be a negative rotation.

Although the both the terms vectors and phasors are used to describe a rotating line that itself has both magnitude and direction, the main difference between the two is that a vectors magnitude is the "peak value" of the sinusoid while a phasors magnitude is the "rms value" of the sinusoid. In both cases the phase angle and direction remains the same.

The phase of an alternating quantity at any instant in time can be represented by a phasor diagram, so phasor diagrams can be thought of as "functions of time". A complete sine wave can be constructed by a single vector rotating at an angular velocity of $\omega = 2\pi f$, where f is the frequency of the waveform. Then a Phasor is a quantity that has both "Magnitude" and "Direction".

Generally, when constructing a phasor diagram, angular velocity of a sine wave is always assumed to be: ω in rad/sec.

Phasor Diagram of a Sinusoidal Waveform

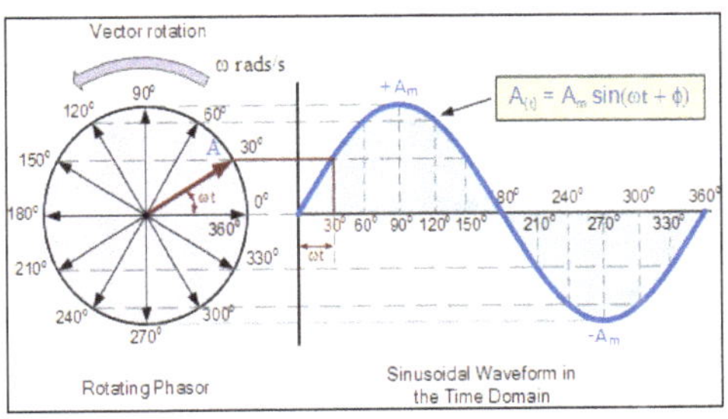

As the single vector rotates in an anti-clockwise direction, its tip at point A will rotate one complete revolution of 360° or 2π representing one complete cycle. If the length of its moving tip is transferred at different angular intervals in time to a graph as shown above, a sinusoidal waveform would be drawn starting at the left with zero time. Each position along the horizontal axis indicates the time that has elapsed since zero time, $t = 0$. When the vector is horizontal the tip of the vector represents the angles at 0°, 180° and at 360°.

Likewise, when the tip of the vector is vertical it represents the positive peak value, (+Am) at 90° or π/2 and the negative peak value, (-Am) at 270° or 3π/2. Then the time axis of the waveform represents the angle either in degrees or radians through which the phasor has moved. So we can say that a phasor represent a scaled voltage or current value of a rotating vector which is "frozen" at some point in time, (t) and in our example above, this is at an angle of 30°.

Sometimes when we are analysing alternating waveforms we may need to know the position of the phasor, representing the Alternating Quantity at some particular instant in time especially when we want to compare two different waveforms on the same axis. For example, voltage and current. We have assumed in the waveform above that the waveform starts at time t = 0 with a corresponding phase angle in either degrees or radians.

But if a second waveform starts to the left or to the right of this zero point or we want to represent in phasor notation the relationship between the two waveforms then we will need to take into account this phase difference, Φ of the waveform.

Phase Difference of a Sinusoidal Waveform

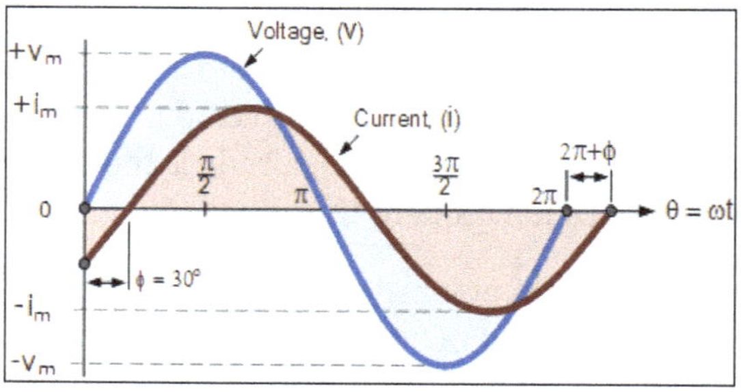

The generalised mathematical expression to define these two sinusoidal quantities will be written as:

$$v_{(t)} = V_m \sin(\omega t)$$
$$i_{(t)} = I_m \sin(\omega t - \phi)$$

The current, i is lagging the voltage, v by angle Φ and in our example above this is 30°. So the difference between the two phasors representing the two sinusoidal quantities is angle Φ and the resulting phasor diagram will be.

Phasor Diagram of a Sinusoidal Waveform

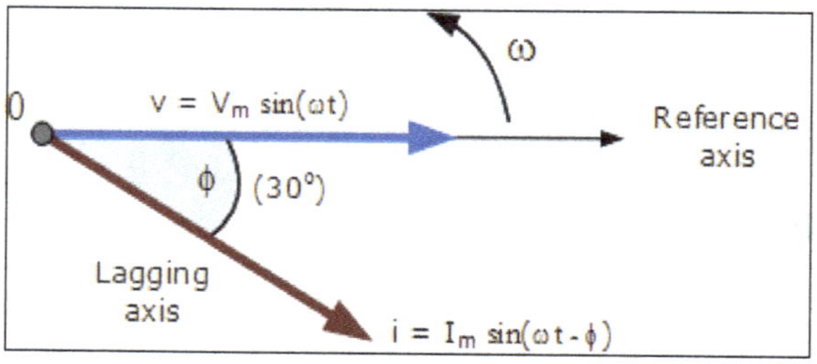

The phasor diagram is drawn corresponding to time zero (t=0) on the horizontal axis. The lengths of the phasors are proportional to the values of the voltage, (V) and the current, (I) at the instant in time that the phasor diagram is drawn. The current phasor lags the voltage phasor by the angle, Φ, as the two phasors rotate in an *anticlockwise*direction as stated earlier, therefore the angle, Φ is also measured in the same anticlockwise direction.

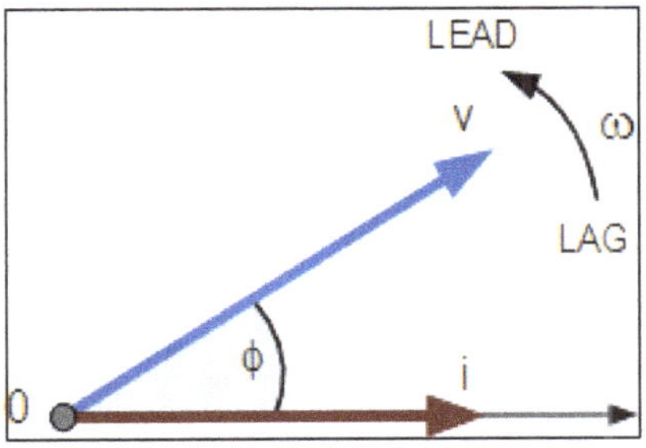

If however, the waveforms are frozen at time, t = 30°, the corresponding phasor diagram would look like the one shown on the right. Once again the current phasor lags behind the voltage phasor as the two waveforms are of the same frequency.

However, as the current waveform is now crossing the horizontal zero axis line at this instant in time we can use the current phasor as our new reference and correctly say that the voltage phasor is "leading" the current phasor by angle, Φ. Either way, one phasor is designated as the *reference* phasor and all the other phasors will be either leading or lagging with respect to this reference.

Phasor Addition

Sometimes it is necessary when studying sinusoids to add together two alternating waveforms, for example in an AC series circuit, that are not in-phase with each other. If they are in-phase that is, there is no phase shift then they can be added together in the same way as DC values to find the algebraic sum of the two vectors. For example, if two voltages of say 50 volts and 25 volts respectively are together "in-phase", they will add or sum together to form one voltage of 75 volts (50 + 25).

If however, they are not in-phase that is, they do not have identical directions or starting point then the phase angle between them needs to be taken into account so they are added together using phasor diagrams to determine their Resultant Phasor or Vector Sum by using the *parallelogram law*.

Consider two AC voltages, V_1 having a peak voltage of 20 volts, and V_2 having a peak voltage of 30 volts where V_1 leads V_2 by 60°. The total voltage, V_T of the two voltages can be found by firstly drawing a phasor diagram representing the two vectors and then constructing a parallelogram in which two of the sides are the voltages, V_1 and V_2 as shown below.

Phasor Addition of Two Phasors

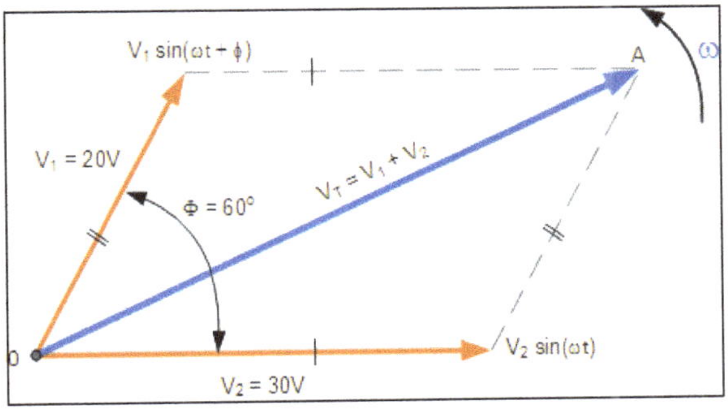

By drawing out the two phasors to scale onto graph paper, their phasor sum $V_1 + V_2$ can be easily found by measuring the length of the diagonal line, known as the "resultant r-vector", from the zero point to the intersection of the construction lines 0-A. The downside of this graphical method is that it is time consuming when drawing the phasors to scale.

Also, while this graphical method gives an answer which is accurate enough for most purposes, it may produce an error if not drawn accurately or correctly to scale. Then one way to ensure that the correct answer is always obtained is by an analytical method.

Mathematically we can add the two voltages together by firstly finding their "vertical" and "horizontal" directions, and from this we can then calculate both the "vertical" and "horizontal" components for the resultant "r vector", V_T. This analytical method which uses the cosine and sine rule to find this resultant value is commonly called the Rectangular Form.

In the rectangular form, the phasor is divided up into a real part, x and an imaginary part, y forming the generalised expression Z = x ± jy. This then gives us a mathematical expression that represents both the magnitude and the phase of the sinusoidal voltage as:

Complex Sinusoid

$$V_m = \cos(\phi) + j V_m (\sin\phi)$$

So the addition of two vectors, A and B using the previous generalised expression is as follows:

$$A = x + jy \quad and \quad B = w + jz$$
$$A + B = (x + w) + j(y + z)$$

Phasor Addition using Rectangular Form

Voltage, V_2 of 30 volts points in the reference direction along the horizontal zero axis, then it has a horizontal component but no vertical component as follows:

- Horizontal Component = 30 cos 0° = 30 volts;
- Vertical Component = 30 sin 0° = 0 volts.

This then gives us the rectangular expression for voltage V_2 of: 30 + j0

Voltage, V_1 of 20 volts leads voltage, V_2 by 60°, then it has both horizontal and vertical components as follows:

- Horizontal Component = 20 cos 60° = 20 x 0.5 = 10 volts;
- Vertical Component = 20 sin 60° = 20 x 0.866 = 17.32 volts.

This then gives us the rectangular expression for voltage V_1 = 10 + j17.32

The resultant voltage, V_T is found by adding together the horizontal and vertical components as follows:

$$V_{Horizontal} = \text{sum of real parts of } V_1 \text{ and } V_2 = 30 + 10 = 40 \text{ volts}$$

$V_{Vertical}$ = sum of imaginary parts of V_1 and V_2 = 0 + 17.32 = 17.32 volts.

Now that both the real and imaginary values have been found the magnitude of voltage, V_T is determined by simply using Pythagoras's Theorem for a 90° triangle as follows:

$$V_T = \sqrt{\begin{pmatrix}\text{Real or Horizontal}\\ \text{Component}\end{pmatrix}^2 + \begin{pmatrix}\text{Imaginary or Vertical}\\ \text{Component}\end{pmatrix}^2}$$

$$V_T = \sqrt{40^2 + 17.32^2}$$

$$\therefore V_T = 43.6 volts$$

Then the resulting phasor diagram will be,

Resultant Value of V_T

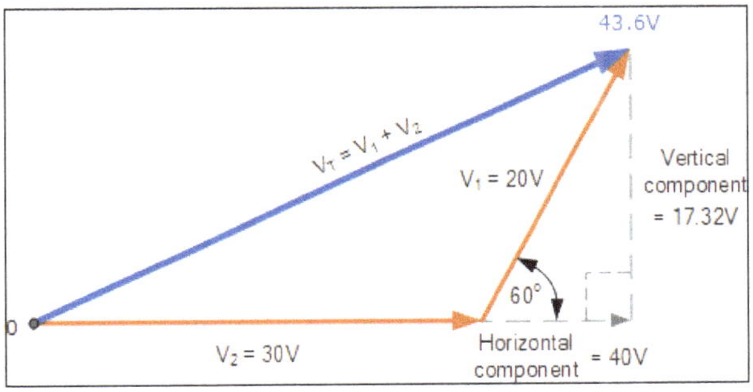

Phasor Subtraction

Phasor subtraction is very similar to the above rectangular method of addition, except this time the vector difference is the other diagonal of the parallelogram between the two voltages of V_1 and V_2 as shown.

Vector Subtraction of Two Phasors

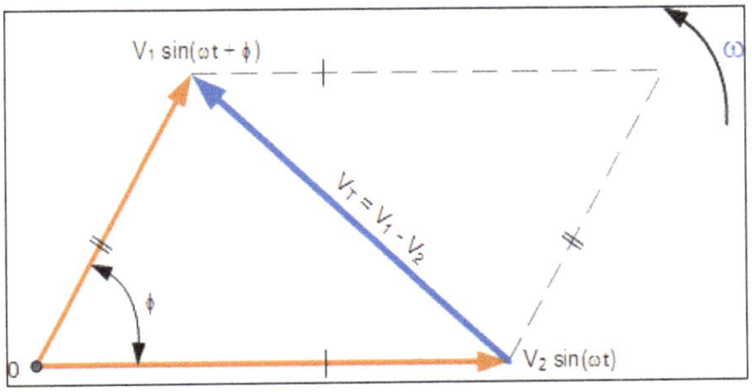

This time instead of "adding" together both the horizontal and vertical components we take them away, subtraction.

$$A = x + jy \quad and \quad B = w + jz$$

$$A - B = (x - w) + j(y - z)$$

The 3-phase Phasor Diagram

Previously we have only looked at single-phase AC waveforms where a single multi-turn coil rotates within a magnetic field. But if three identical coils each with the same number of coil turns are placed at an electrical angle of 120° to each other on the same rotor shaft, a three-phase voltage supply would be generated.

A balanced three-phase voltage supply consists of three individual sinusoidal voltages that are all equal in magnitude and frequency but are out-of-phase with each other by exactly 120° electrical degrees.

Standard practice is to colour code the three phases as Red, Yellow and Blue to identify each individual phase with the red phase as the reference phase. The normal sequence of rotation for a three phase supply is Red followed by Yellow followed by Blue, (R, Y, B).

As with the single-phase phasors above, the phasors representing a three-phase system also rotate in an anti-clockwise direction around a central point as indicated by the arrow marked ω in rad/s. The phasors for a three-phase balanced star or delta connected system are shown below.

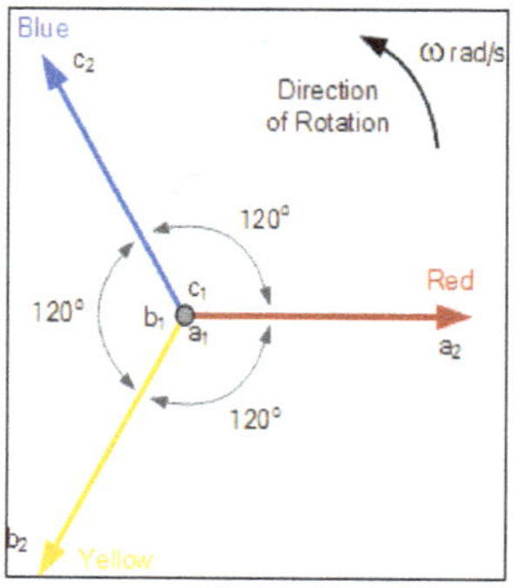

The phase voltages are all equal in magnitude but only differ in their phase angle. The three windings of the coils are connected together at points, a_1, b_1 and c_1 to produce a common neutral connection for the three individual phases. Then if the red phase is taken as the reference phase each individual phase voltage can be defined with respect to the common neutral as.

Three-phase Voltage Equations

$$\text{Red Phase: } V_{RN} = V_m \sin\theta$$

$$\text{Yellow Phase: } V_{YN} = V_m \sin\left(\theta - 120^0\right)$$

$$\text{Blue Phase } V_{BN} = V_m \sin\left(\theta - 240^0\right)$$

or

$$V_{RN} = V_m \sin\left(\theta + 120^0\right)$$

If the red phase voltage, V_{RN} is taken as the reference voltage as stated earlier then the phase sequence will be R – Y – B so the voltage in the yellow phase lags V_{RN} by 120°, and the voltage in the blue phase lags V_{YN} also by 120°. But we can also say the blue phase voltage, V_{BN} leads the red phase voltage, V_{RN} by 120°.

One final point about a three-phase system. As the three individual sinusoidal voltages have a fixed relationship between each other of 120° they are said to be "balanced" therefore, in a set of balanced three phase voltages their phasor sum will always be zero as: $V_a + V_b + V_c = 0$,

Electrical Impedance

Electrical impedance (also known as "impedance" in short) is an addition of the definition of resistance to analternating current(AC). Which means that impedance includes both resistance (opposition of the electric current that causes heat) and reactance (a measure of such an oppositional current alternates) – in detail, the opposition adjacent to the electric currents. In the direct current (DC), electrical impedance is same as a resistance, except that it does not hold true in AC circuits.

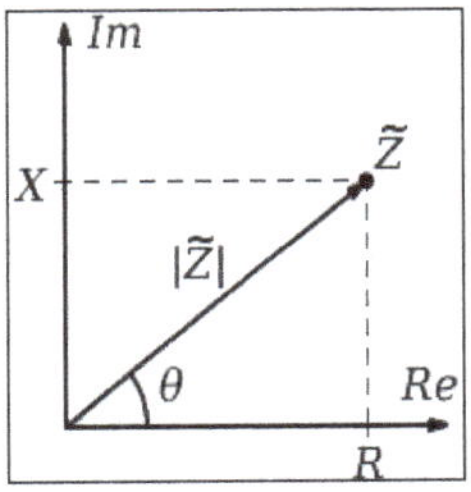

Electrical Impedance.

Impedance can also be dissimilar from resistance when a DC circuit changes flow in one way or other- similar to the opening and closing of an electrical switch, as is observed in the computers when they open and close switches to represent ones and zeros (binary language). The opposite of an impedance is admittance, which is the measure of the allowance of current. The figure to the left is a complex impedance plane, in which impedance is represented by a Z, resistance is depicted as R, and the reactance is depicted with X.

Electrical Impedance Tomography (EIT)

The fundamental principle of electrical impedance tomography (EIT) is akin to electrical resistance tomography (ERT) such that several measurements at the periphery of a process vessel or tube are taken and combined to give information on the electrical properties of the process volume.

Electrical Impedance Tomography.

Electrical Impedance Tomography (EIT) is a non-invasive medical imaging method in which a figure of the conductivity or permitivity of a part of the body is incidental from the surface electrode measurements. Electrical conductivity depends on the free ion content and differs significantly between different biological tissues (absolute EIT) or dissimilar practical states of one and other similar tissues or organs (relative or functional EIT). Majority of EIT systems apply little irregular currents at a single frequency; however, some EIT systems use various frequencies to better discriminate between usual and suspected abnormal tissue within the same organ (multifrequency-EIT or electrical impedance spectroscopy).

Complex Impedance

A resistor with a value of R has an impedance of R ohms, a real number. An ideal inductor has a complex impedance of,

$$Z = j2\pi L$$

Where, 'f' is the frequency in Hertz and L is the inductance in Henries. It is imaginary because an a ideal inductor can simply store and release electrical energy. It can't dissipate it as heat like a resistor. Similarly, an ideal capacitor has a complex impedance of,

$$Z = \frac{-j}{2\pi fc}$$

Where 'C' is the capacitance in farads.

Use of Complex Impedance

The behavior of the impedance of an AC circuit with various components quickly becomes unmanageable if sines and cosines are used to present the voltages and current. A mathematical build which eases the complexity use of complex exponential functions. The necessary parts of the strategy are as follows.

Math relationship underlaying the technique,

$$ej\omega t = cos\omega t + sin\omega t$$

The real part of a complex exponential function can be used to represent an AC voltage or current.

$$V = Vm\ COS\omega t$$
$$I = Im\ COS\ (\omega t - \varphi)$$

The impedance can then be expressed as a complex exponential,

$$Z = Vm /\ Im\ e - j\O = R + jX$$

The impedance of the individual circuit elements can then be expressed as pure real or imaginary numbers.

$$R\ - j / \omega c\ j\omega L$$

Complex Impedance for RL and RC

Using complex impedance is a significant technique for handling multi-component AC circuits. If a complex plane is used with resistance along the real axis, then the reactance of a capacitor and inductor are treated as imaginary numbers. For series combinations of the components such as RL and RC combinations, the component values are added as if they were components of a vector. Shown now is the Cartesian form of the complex impedance. They can also be written in polar form. Impedances in combination circuits like the RLC parallel circuit.

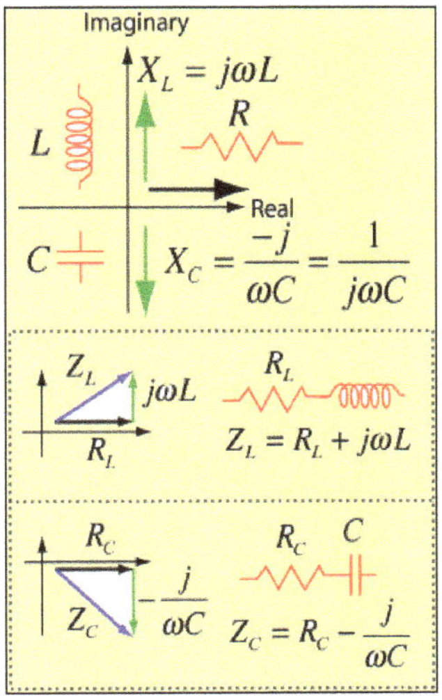

Complex Impedance for RL and RC.

Resistance and Reactance

Resistance is fundamentally a friction against the motion of electrons. It is there in all conductors to some extent (except superconductors!), and most notably in resistors. When alternating current goes through a resistance, a voltage drop is formed that is in-phase with the current. Resistance is mathematically symbolized by the letter "R" and is measured in the unit of ohms (Ω).

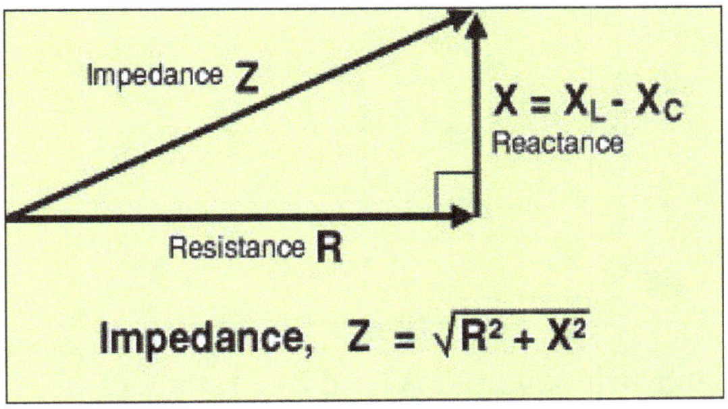

Resistance and Reactance Circuit.

Reactance is essentially inactive against the motion of electrons. It is present anywhere electric or magnetic fields are developed in proportion to an applied voltage or current, correspondingly; but most notably in capacitors and inductors. When alternating current goes through a pure reactance, a voltage drop is produced, which is 900 out of phase with the current. Reactance is a mathematically symbolized by the letter "X" and is measured in the unit of Ohms (Ω).

Network Analysis

Network Analysis for electric circuits are the different useful techniques related to several currents, emfs, and resistance voltages in such circuit. This is somewhat the collection of techniques of finding the voltages and currents in every component of the network.

Nodal Analysis

Nodal analysis is a method that provides a general procedure for analyzing circuits using node voltages as the circuit variables. Nodal Analysis is also called the Node-Voltage Method.

Some features of nodal analysis are as follows:

- Nodal Analysis is based on the application of the Kirchhoff's Current Law (KCL).
- Having 'n' nodes there will be 'n-1' simultaneous equations to solve.
- Solving 'n-1' equations all the nodes voltages can be obtained.
- The number of non reference nodes is equal to the number of Nodal equations that can be obtained.

Types of Nodes in Nodal Analysis

- Non Reference Node: It is a node which has a definite Node Voltage. e.g. Here Node 1 and Node 2 are the Non Reference nodes.
- Reference Node: It is a node which acts a reference point to all the other node. It is also called the Datum Node.

Types of Reference Nodes

1. Chassis Ground: This type of reference node acts a common node for more than one circuits.

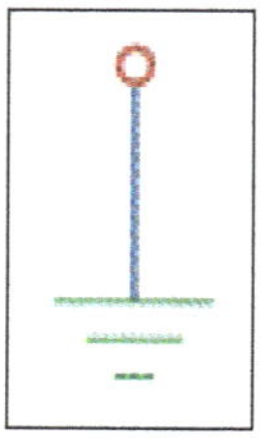

2. Earth Ground: When earth potential is used as a reference in any circuit then this type of reference node is called Earth Ground.

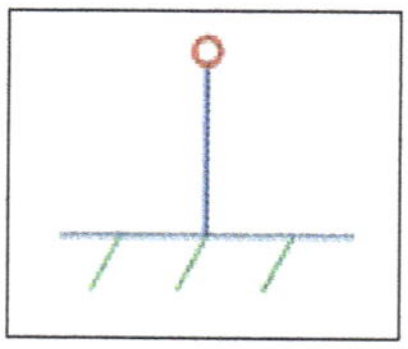

Solving of Circuit using Nodal Analysis

Basic Steps used in Nodal Analysis

- Select a node as the reference node. Assign voltages V_1, V_2... V_{n-1} to the remaining nodes. The voltages are referenced with respect to the reference node.
- Apply KCL to each of the non reference nodes.
- Use Ohm's law to express the branch currents in terms of node voltages.

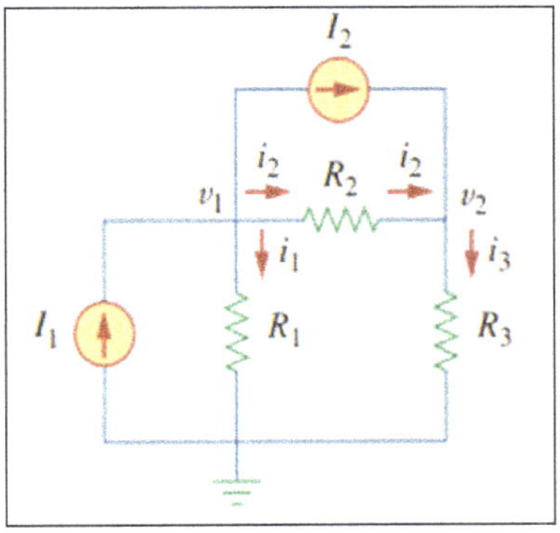

Node always assumes that current flows from a higher potential to a lower potential in resistor. Hence, current is expressed as follows,

$$I = \frac{V_{high} - V_{low}}{R}$$

- After the application of Ohm's Law get the 'n-1' node equations in terms of node voltages and resistances.
- Solve 'n-1' node equations for the values of node voltages and get the required node Voltages as result.

Nodal Analysis with Current Sources

Nodal analysis with current sources is very easy and it is discussed with a example below.

Example: Calculate Node Voltages in following circuit,

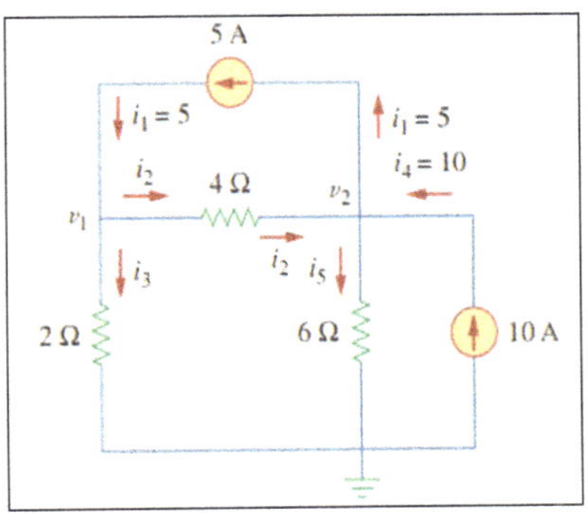

In the following circuit we have 3 nodes from which one is reference node and other two are non reference nodes - Node 1 and Node 2.

Step I. Assign the nodes voltages as v_1 and v_2 and also mark the directions of branch currents with respect to the reference nodes.

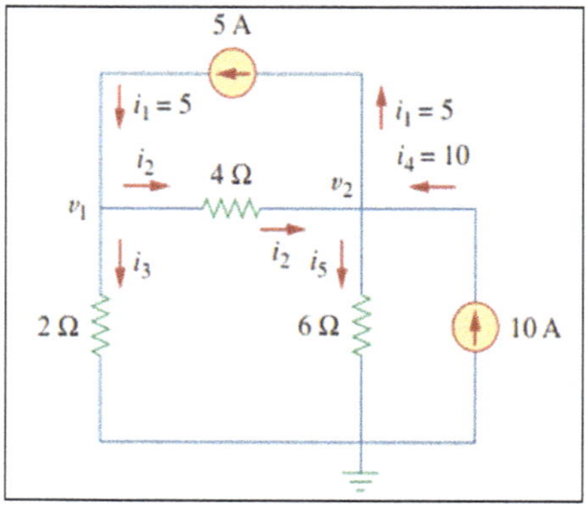

Step II. Apply KCL to Nodes 1 and 2.

KCL at Node 1,

$$i_1 = i_2 + i_3$$

KCL at Node 2,

$$i_2 + i_4 = i_1 + i_5$$

Step III. Apply Ohm's Law to KCL equations,

- Ohm's law to KCL equation at Node 1,

$$i_1 = i_2 + i_3 \Rightarrow 5 = \frac{v_1 - v_2}{4} + \frac{v_2 - 0}{4}$$

Simplifying the above equation we get,

$$3v_1 - v_2 = 20$$

- Now, Ohm's Law to KCL equation at Node 2,

$$i_2 + i_4 = i_1 + i_5 \Rightarrow \frac{v_1 - v_2}{4} + 10 = 5 + \frac{v_2 - 0}{6}$$

Simplifying the above equation we get,

$$-3v_1 + 5v_2 = 60$$

Step IV. Now solve the above two equations to get the values of v_1 and v_2.

Using elimination method,

$$3v_1 - v_2 = 20$$
$$-3v_1 + 5v_2 = 60$$
$$\Rightarrow 4v_2 = 80 \Rightarrow v_2 = 20\,Volts$$

And substituting value v_2 = 20 Volts in equation mentioned before the above one, we get,

$$3v_1 - 20 = 20 \Rightarrow v_1 = \frac{40}{3} = 13.333\,Volts$$

Hence node voltages are as v1 = 13.33 Volts and v2 = 20 Volts.

Nodal Analysis with Voltage Sources

Case I. If a voltage source is connected between the reference node and a non reference node, we simply set the voltage at the non-reference node equal to the voltage of the voltage source and its analysis can be done as we done with current sources. v_1 = 10 Volts.

Case II. If the voltage source is between the two non reference nodes then it forms a supernode whose analysis is done as following:

Supernode Analysis

Whenever a voltage source (Independent or Dependent) is connected between the two non reference nodes then these two nodes form a generalized node called the Super node. So, Super node can be regarded as a surface enclosing the voltage source and its two nodes.

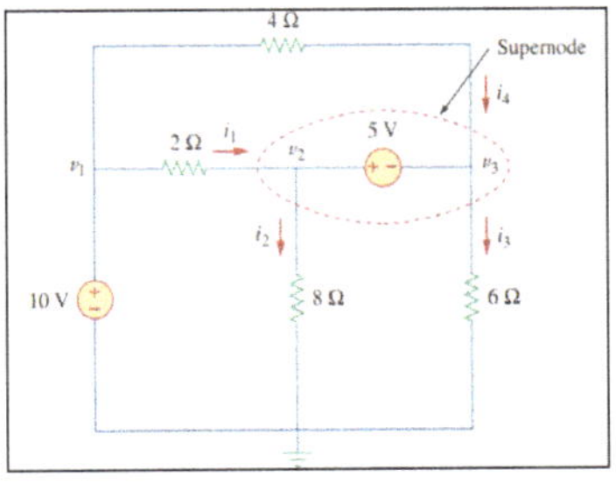

In the above figure source is connected between two non reference nodes Node - 2 and Node - 3. So here Node - 2 and Node - 3 form the Super node.

Properties of Supernode

- Always the difference between the voltage of two non reference nodes is known at Supernode.
- A supernode has no voltage of its own.
- A supernode requires application of both KCL and KVL to solve it.
- Any element can be connected in parallel with the voltage source forming the supernode.
- A Supernode satisfies the KCL as like a simple node.

Solving Any Circuit Containing Supernode

Let's take a example to understand how to solve circuit containing Supernode.

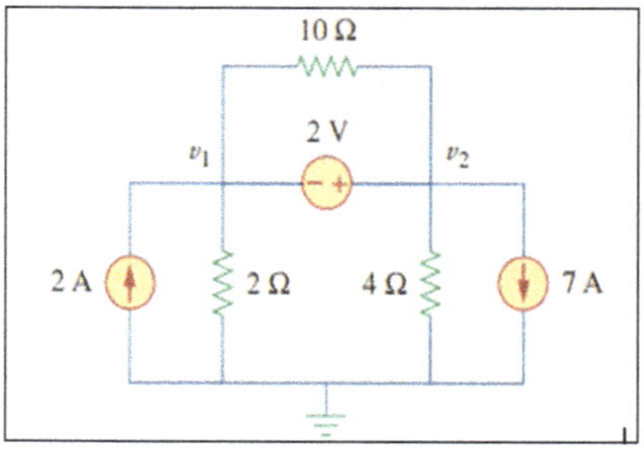

Here 2V voltage source is connected between Node-1 and Node-2 and it forms a Supernode with a 10Ω resistor in parallel.

Note - Any element connected in parallel with the voltage source forming Super node doesn't make any difference because $v_2 - v_1 = 2V$ always whatever may be the value of resistor. Thus 10 Ω can be removed and circuit is redrawn and applying KCL to the supernode as shown in figure gives,

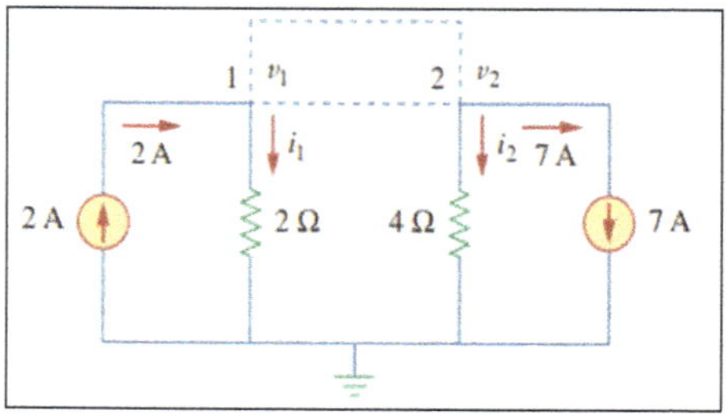

Expressing and in terms of the node voltages.

$$2 = \frac{v_1 - 0}{2} + \frac{v_2 - 0}{4} + 7 \Rightarrow 8 = 2v_1 + v_2 + 28 \Rightarrow v_2 = -2v_1 - 20$$

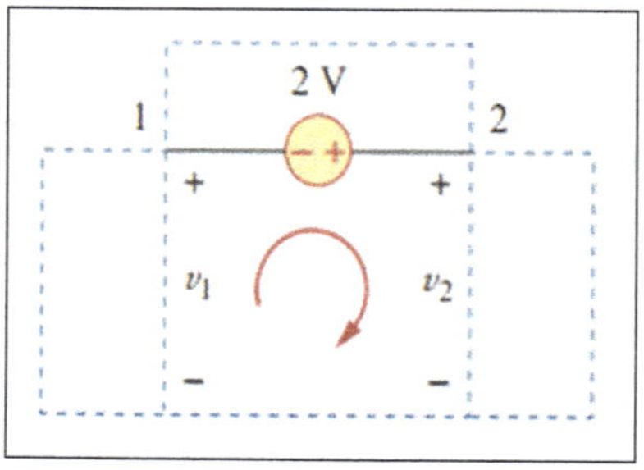

$$v_1 + 2 - v_2 = 0 \Rightarrow v_2 = v_1 + 2$$

From above two equations we can write as,

$$v_2 = v_1 + 2 = -2v_1 - 20 \Rightarrow 3v_1 = -22$$
$$\Rightarrow v_1 = -7.333V \;\&\; v_2 = v_1 + 2 = -7.333 + 2 = -5.333V$$

Hence, v_1 = - 7.333V and v_2 = - 5.333V which is required answer.

Mesh Analysis

Mesh Current Analysis is a technique used to find the currents circulating around a loop or mesh with in any closed path of a circuit.

Mesh Current Analysis Circuit

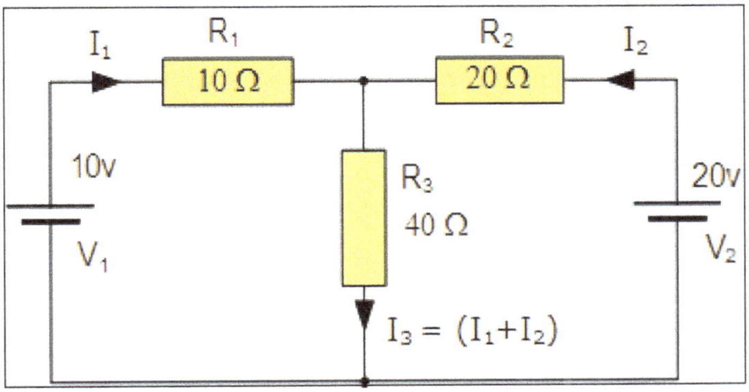

One simple method of reducing the amount of math's involved is to analyse the circuit using Kirchhoff's Current Law equations to determine the currents, I_1 and I_2 flowing in the two resistors. Then there is no need to calculate the current I_3 as its just the sum of I_1 and I_2. So Kirchhoff's second voltage law simply becomes:

$$10 = 50I_1 + 40I_2$$

$$20 = 40I_1 + 60I_2$$

therefore, one line of math's calculation have been saved.

Mesh Current Analysis

An easier method of solving the above circuit is by using Mesh Current Analysis or Loop Analysis which is also sometimes called Maxwell´s Circulating Currents method. Instead of labelling the branch currents we need to label each "closed loop" with a circulating current.

As a general rule of thumb, only label inside loops in a clockwise direction with circulating currents as the aim is to cover all the elements of the circuit at least once. Any required branch current may be found from the appropriate loop or mesh currents as before using Kirchhoff´s method.

For example: $i_1 = I_1$, $i_2 = -I_2$ and $I_3 = I_1 - I_2$

We now write Kirchhoff's voltage law equation in the same way as before to solve them but the advantage of this method is that it ensures that the information obtained from the circuit equations is the minimum required to solve the circuit as the information is more general and can easily be put into a matrix form.

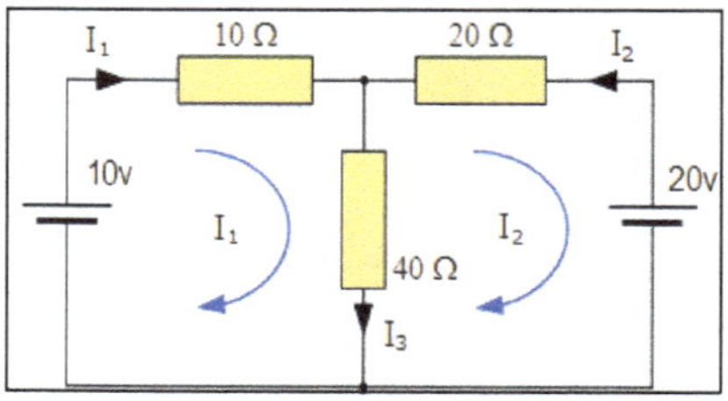

$10=50I_1-40I_2$

$-20=.40I_1+60I_2$

or [V]=[R][I]

These equations can be solved quite quickly by using a single mesh impedance matrix Z. Each element ON the principal diagonal will be "positive" and is the total impedance of each mesh. Where as, each element OFF the principal diagonal will either be "zero" or "negative" and represents the circuit element connecting all the appropriate meshes.

First we need to understand that when dealing with matrices, for the division of two matrices it is the same as multiplying one matrix by the inverse of the other as shown.

$$[V]=[I]\times[R] \ or \ [R]\times[I]=[V]$$

$$\begin{bmatrix} 50 & -40 \\ -40 & 60 \end{bmatrix} \times \begin{bmatrix} I_1 \\ I_2 \end{bmatrix} = \begin{bmatrix} 10 \\ -20 \end{bmatrix}$$

$$I = \frac{V}{R} = R^{-1} \times V$$

$$Inverse \ of \ R = \begin{bmatrix} 60 & 40 \\ 40 & 50 \end{bmatrix}$$

$$|R| = (60 \times 50) - (40 \times 40) = 1400$$

$$\therefore R^{-1} = \frac{1}{1400} \begin{bmatrix} 60 & 40 \\ 40 & 50 \end{bmatrix}$$

Having found the inverse of R, as V/R is the same as V x R^{-1}, we can now use it to find the two circulating currents.

$$[I] = \left[R^{-1}\right] \times [V]$$

$$\begin{bmatrix} I_1 \\ I_2 \end{bmatrix} = \frac{1}{1400} \begin{bmatrix} 60 & 40 \\ 40 & 50 \end{bmatrix} \times \begin{bmatrix} 10 \\ -20 \end{bmatrix}$$

$$I_1 = \frac{(60\times10)+(40\times-20)}{1400} = \frac{-200}{1400} = -0.143A$$

$$I_2 = \frac{(40\times10)+(50\times-20)}{1400} = \frac{-600}{1400} = -0.429A$$

Where,

[V] gives the total battery voltage for loop 1 and then loop 2.

[I] states the names of the loop currents which we are trying to find.

[R] is the resistance matrix.

[R^{-1}] is the inverse of the [R] matrix.

and this gives I_1 as -0.143 Amps and I_2 as -0.429 Amps.

$I_3 = I_1 - I_2$

The combined current of I_3 is therefore given as : -0.143 – (-0.429) = 0.286 Amps

which is the same value of 0.286 amps.

Ohm's Law

Ohm's Law states that the current flowing in a circuit is directly proportional to the applied potential difference and inversely proportional to the resistance in the circuit.

In other words by doubling the voltage across a circuit the current will also double. However if the resistance is doubled the current will fall by half.

In this mathematical relationship the unit of resistance is measured in Ohms.

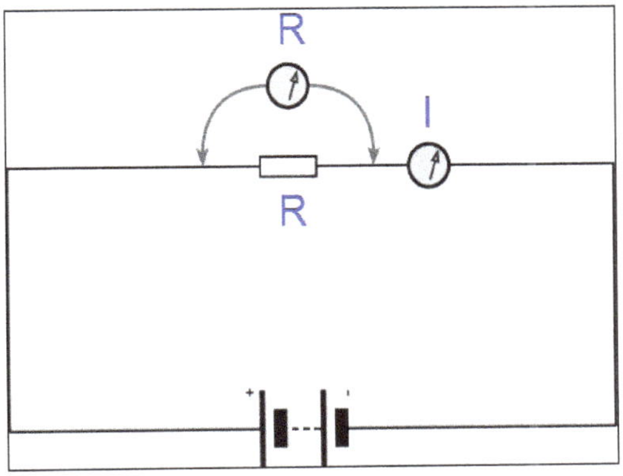

Ohm's Law Formula

The Ohm's Law formula or equation is very straightforward.

Ohm's law can be expressed in a mathematical form:

$$V = IR$$

Where,

V = voltage expressed in Volts.

I = current expressed in Amps.

R = resistance expressed in Ohms.

The formula can be manipulated so that if any two quantities are known the third can be calculated.

$$I = \frac{V}{R}$$

$$R = \frac{V}{I}$$

Ohm's Law Triangle

To help remember the formula it is possible to use a triangle with one side horizontal and the peak at the top like a pyramid. This is sometimes known as the Ohm's law triangle.

In the top corner of the Ohm's law triangle is the letter V, in the left hand corner, the letter I, and in the right hand bottom corner, R.

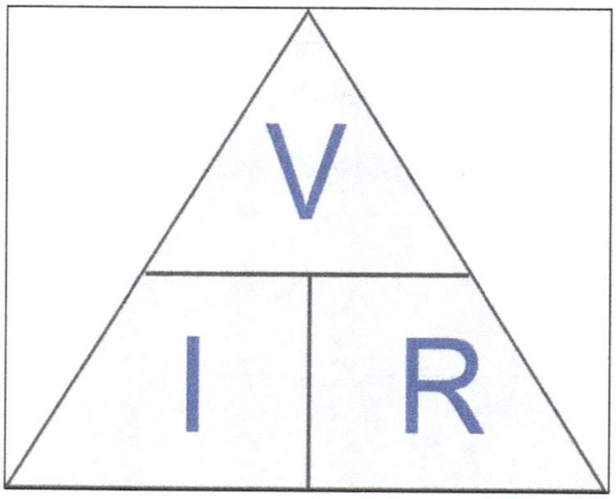

To use the triangle cover up the unknown quantity and then and then calculate it from the other two. If they are in line they are multiplied, but if one is on top of the other then they should be divided. In other words if current has to be calculated the voltage is divided by the resistance i.e. V/R and so forth.

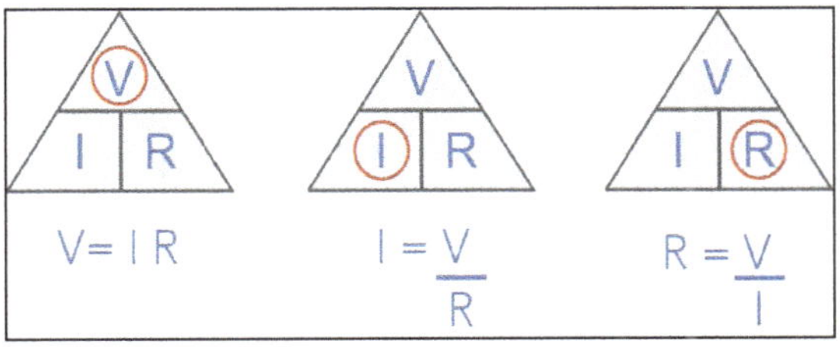

If the voltage has to be calculated then it is found by multiplying the current by the resistance i.e. I x R.

Ohm's Law Calculation Example

If a voltage of 10 volts is placed across a 500 ohm resistor determine the amount of current that will flow.

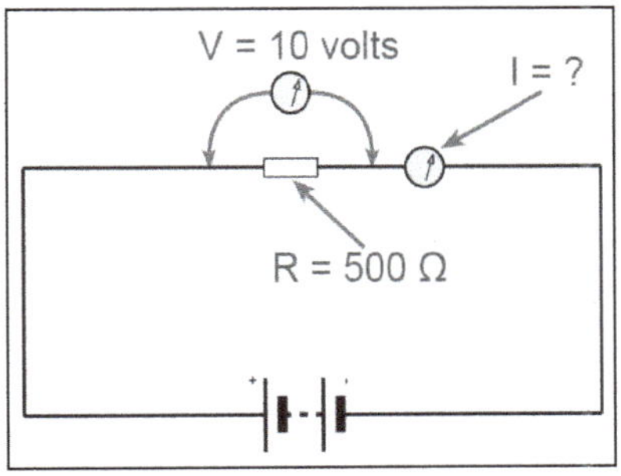

Looking at the Ohm's Law triangle the current is the unknown leaving the voltage and resistance as the known values.

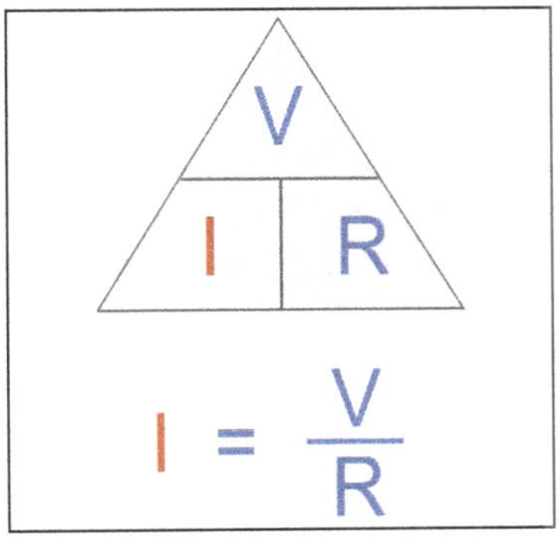

In this way the current is found by dividing the voltage by the resistance.

$$I = \frac{V}{R} = \frac{10}{500} = 0.02A = 20mA$$

Linear Graph

It can be seen that if the voltage and current were plotted for a fixed resistor or a length of wire, etc, there would be a linear curve.

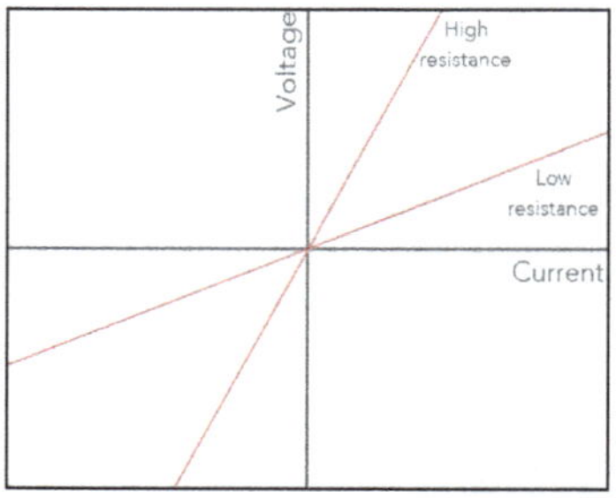

Plot of voltage and current for a linear resistance.

It can be seen that doubling the voltage doubles the current that passes through the particular circuit element.

On the graph, there are two lines, one for a higher resistance - this one requires more voltage to be applied for a given current flowing. Accordingly this must have a higher resistance. Conversely the curve for the lower resistance shows a component that requires a lower voltage to be applied for a given current.

Non-linear Components

Ohm's law in its basic form where a doubling in voltage results in a doubling in current applies to linear components like ordinary resistors. Some components like diodes have non-linear curves where the resistance is affected by the applied voltage.

Ohm's Law is one of the most basic concepts within electrical and electronic engineering. The concept of an item having a certain resistance which determines the amount of current that flows through it for a certain voltage is key to the operation virtually all circuits.

Ohm's Law Pie Chart

To help us understand the the relationship between the various values a little further, we can take all of the Ohm's Law equations from above for finding Voltage, Current, Resistance and of course Power and condense them into a simple Ohm's Law pie chart for use in AC and DC circuits and calculations as shown,

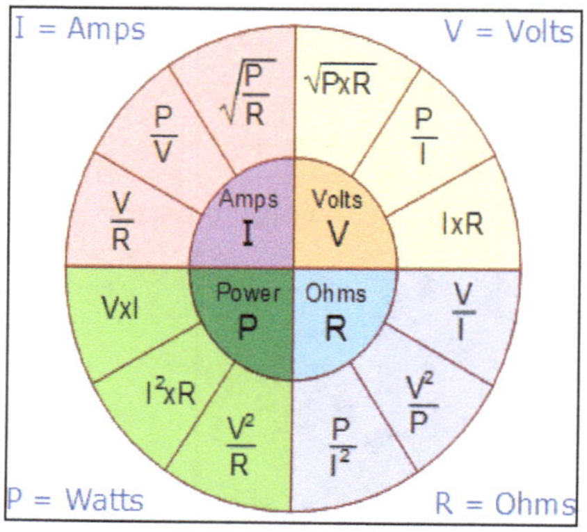

As well as using the Ohm's Law Pie Chart shown above, we can also put the individual Ohm's Law

equations into a simple matrix table as shown for easy reference when calculating an unknown value.

Ohm's Law Matrix Table

Ohms Law Formulas				
Known Values	Resistance (R)	Current (I)	Voltage (V)	Power (P)
Current & Resistance	---	---	$V = I \times R$	$P = I^2 \times R$
Voltage & Current	$R = \frac{V}{I}$	---	---	$P = V \times I$
Power & Current	$R = \frac{P}{I^2}$	---	$V = \frac{P}{I}$	---
Voltage & Resistance	---	$I = \frac{V}{R}$	---	$P = \frac{V^2}{R}$
Power & Resistance	---	$I = \sqrt{\frac{P}{R}}$	$V = \sqrt{P \times R}$	---
Voltage & Power	$R = \frac{V^2}{P}$	$I = \frac{P}{V}$	---	---

Ohm's Law Example

For the circuit shown below find the Voltage (V), the Current (I), the Resistance (R) and the Power (P).

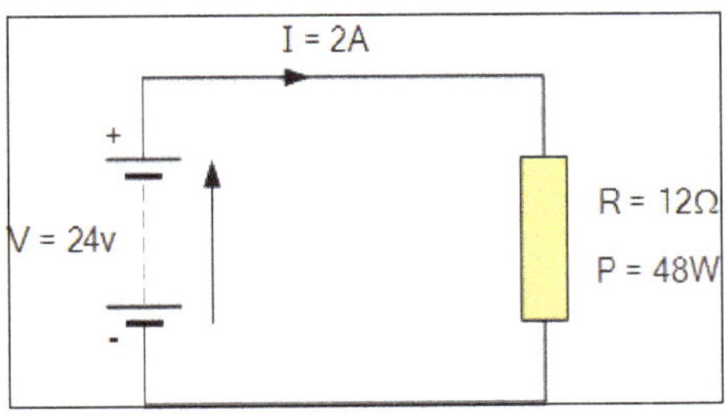

Voltage [V = I x R] = 2 x 12Ω = 24V

Current [I = V ÷ R] = 24 ÷ 12Ω = 2A

Resistance [R = V ÷ I] = 24 ÷ 2 = 12 Ω

Power [P = V x I] = 24 x 2 = 48W

Power within an electrical circuit is only present when BOTH voltage and current are present. For example, in an open-circuit condition, voltage is present but there is no current flow I = 0 (zero), therefore V*0 is 0 so the power dissipated within the circuit must also be 0. Likewise, if we have a short-circuit condition, current flow is present but there is no voltage V = 0, therefore 0*I = 0 so again the power dissipated within the circuit is 0.

As electrical power is the product of V*I, the power dissipated in a circuit is the same whether the circuit contains high voltage and low current or low voltage and high current flow. Generally, electrical power is dissipated in the form of Heat (heaters), Mechanical Work such as motors, Energy in the form of radiated (Lamps) or as stored energy (Batteries).

Kirchhoff's Circuit Laws

Kirchhoff's circuit laws are two equations first published by Gustav Kirchhoff in 1845. Fundamentally, they address conservation of energy and charge in the context of electrical circuits.

Although Kirchhoff's Laws can be derived from the equations of James Clerk Maxwell, Maxwell did not publish his set of differential equations (which form the foundation of classical electrodynamics, optics, and electric circuits) until 1861 and 1862. Kirchhoff, rather, used Georg Ohm's work as a foundation for *Kirchhoff's current law (KCL)* and *Kirchhoff's voltage law (KVL).*

Kirchhoff's laws are extremely important to the analysis of closed circuits. Consider, for example, the circuit illustrated in the figure below, consisting of five resistors in a combination of in series and parallel arrangements. Simplification of this circuit to a combination of series and parallel connections is impossible. However, using Kirchhoff's rules, one can analyze the circuit to determine the parameters of this circuit using the values of the resistors (R_1, R_2, R_3, r_1 and r_2). Also of importance in this example is that the values $E_1$and E_2 represent sources of voltage (e.g., batteries).

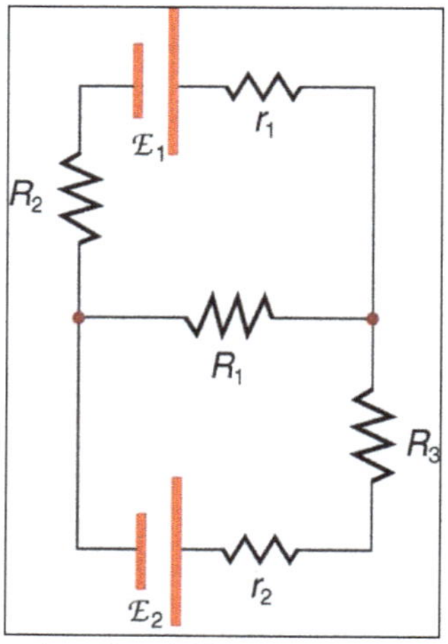

Closed Circuit: To determine all variables (i.e., current and voltage drops across the different resistors) in this circuit, Kirchhoff's rules must be applied.

As a final note, Kirchhoff's laws depend on certain conditions. The voltage law is a simplification of Faraday's law of induction, and is based on the assumption that there is no fluctuating magnetic field within the closed loop. Thus, although this law can be applied to circuits containing resistors and capacitors (as well as other circuit elements), it can only be used as an approximation to the behavior of the circuit when a changing current and therefore magnetic field are involved.

The Junction Rule

Kirchhoff's junction rule states that at any circuit junction, the sum of the currents flowing into and out of that junction are equal.

Kirchhoff's junction rule, also known as Kirchhoff's current law (KCL), Kirchoff's first law, Kirchhoff's point rule, and Kirchhoff's nodal rule, is an application of the principle of conservation of electric charge.

Kirchhoff's junction rule states that at any junction (node) in an electrical circuit, the sum of the currents flowing into that junction is equal to the sum of the currents flowing out of that junction. In other words, given that a current will be positive or negative depending on whether it is flowing towards or away from a junction, the algebraic sum of currents in a network of conductors meeting at a point is equal to zero.

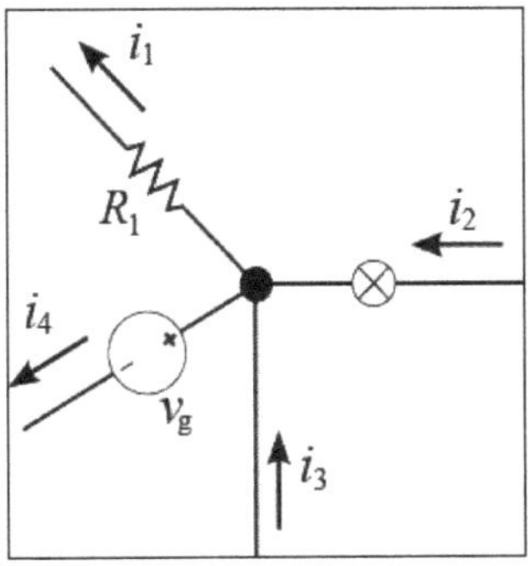

Kirchhoff's Junction Law illustrated as currents flowing into and out of a junction.

Thus, Kirchoff's junction rule can be stated mathematically as a sum of currents (I):

$$\sum_{k=1}^{n} I_k = 0$$

where n is the total number of branches carrying current towards or away from the node.

This law is founded on the conservation of charge (measured in coulombs), which is the product of current (amperes) and time (seconds).

Limitation

Kirchhoff's junction law is limited in its applicability. It holds for all cases in which total electric charge (Q) is constant in the region in consideration. Practically, this is always true so long as the law is applied for a specific point. Over a region, however, charge density may not be constant. Because charge is conserved, the only way this is possible is if there is a flow of charge across the boundary of the region. This flow would be a current, thus violating Kirchhoff's junction law.

The Loop Rule

Kirchhoff's loop rule states that the sum of the emf values in any closed loop is equal to the sum of the potential drops in that loop.

Kirchhoff's loop rule (otherwise known as Kirchhoff's voltage law (KVL), Kirchhoff's mesh rule, Kirchhoff's second law, or Kirchhoff's second rule) is a rule pertaining to circuits, and is based on the principle of conservation of energy.

Conservation of energy—the principle that energy is neither created nor destroyed—is a ubiquitous principle across many studies in physics, including circuits. Applied to circuitry, it is implicit that the directed sum of the electrical potential differences (voltages) around any closed network is equal to zero. In other words, the sum of the electromotive force (emf) values in any closed loop is equal to the sum of the potential drops in that loop (which may come from resistors).

Another equivalent statement is that the algebraic sum of the products of resistances of conductors (and currents in them) in a closed loop is equal to the total electromotive force available in that loop. Mathematically, Kirchhoff's loop rule can be represented as the sum of voltages in a circuit, which is equated with zero:

$$\sum_{k=1}^{n} V_k = 0.$$

Here, V_k is the voltage across element k, and n is the total number of elements in the closed loop circuit. An illustration of such a circuit is shown in figure. In this example, the sum of v_1, v_2, v_3, and v_4 (and v_5 if it is included), is zero.

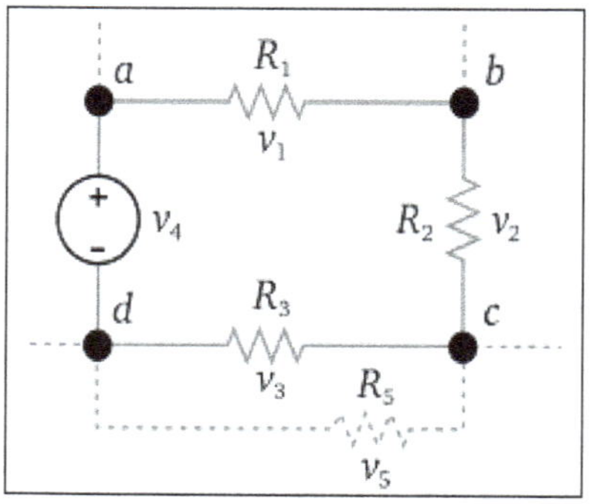

Kirchhoff's Loop Rule: Kirchhoff's loop rule states that the sum of all the voltages around the loop is equal to zero: v1 + v2 + v3 – v4 = 0,

Given that voltage is a measurement of energy per unit charge, Kirchhoff's loop rule is based on the law of conservation of energy which states: the total energy gained per unit charge must equal the amount of energy lost per unit of charge.

Example

The following illustrates the changes in potential in a simple series circuit loop. Kirchhoff's second rule requires emf–Ir–IR$_1$–IR$_2$=0. Rearranged, this is emf=Ir+IR$_1$+IR$_2$, meaning that the emf equals the sum of the IR (voltage) drops in the loop. The emf supplies 18 V, which is reduced to zero by the resistances, with 1 V across the internal resistance, and 12 V and 5 V across the two load resistances, for a total of 18 V.

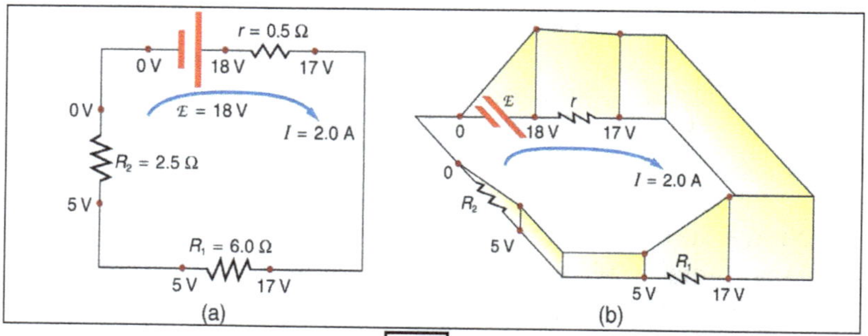

The Loop Rule: An example of Kirchhoff's second rule where the sum of the changes in potential around a closed loop must be zero:

(a) In this standard schematic of a simple series circuit, the emf supplies 18 V, which is reduced to zero by the resistances, with 1 V across the internal resistance, and 12 V and 5 V across the two load resistances, for a total of 18 V.

(b) This perspective view represents the potential as something like a roller coaster, where charge is raised in potential by the emf and lowered by the resistances.

Limitation

Kirchhoff's loop rule is a simplification of Faraday's law of induction, and holds under the assumption that there is no fluctuating magnetic field linking the closed loop. In the presence of a variable magnetic field, electric fields could be induced and emf could be produced, in which case Kirchhoff's loop rule breaks down.

Applications

Kirchhoff's rules can be used to analyze any circuit and modified for those with EMFs, resistors, capacitors and more. Kirchhoff's rules can be used to analyze any circuit by modifying them for those circuits with electromotive forces, resistors, capacitors and more. Practically speaking, however, the rules are only useful for characterizing those circuits that cannot be simplified by combining elements in series and parallel.

Combinations in series and parallel are typically much easier to perform than applying either of Kirchhoff's rules, but Kirchhoff's rules are more broadly applicable and should be used to solve problems involving complex circuits that cannot be simplified by combining circuit elements in series or parallel.

Example of Kirchoff's Rules

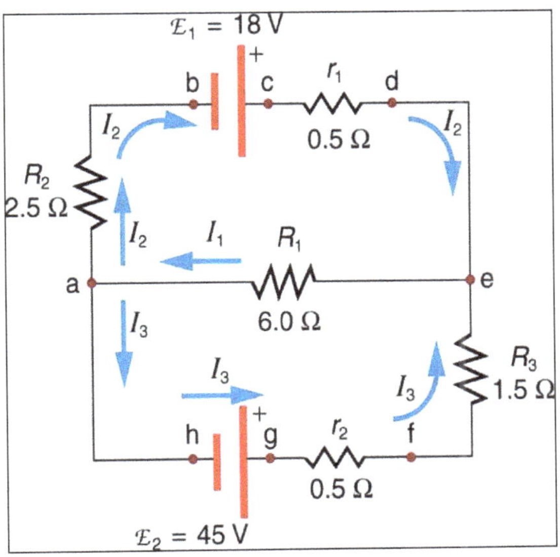

Shows a very complex circuit, but Kirchhoff's loop and junction rules can be applied. To solve the circuit for currents I_1, I_2, and I_3, both rules are necessary:

Kirchhoff's Rules sample problem: This figure shows a very complicated circuit, which can be reduced and solved using Kirchoff's Rules.

Applying Kirchhoff's junction rule at point a, we find:

$$I_1 = I_2 + I_3$$

because I_1 flows into point a, while I_2 and I_3 flow out. The same can be found at point e. We now must solve this equation for each of the three unknown variables, which will require three different equations.

Considering loop abcdea, we can use Kirchhoff's loop rule:

$$-I_2R_2 + emf_1 - I_2r_1 - I_1R_1 = -I_2(R_2) + r_1) + emf_1 - I_1R_1 = 0$$

Substituting values of resistance and emf from the figure diagram and canceling the ampere unit gives:

$$-3I_2 + 18 - 6I_1 = 0$$

This is the second part of a system of three equations that we can use to find all three current values. The last can be found by applying the loop rule to loop aefgha, which gives:

$$I_1R_1 + I_3R_3 + I_3r_2 - emf_2 = I_1R_1 + I_3(R_3 + r_2) - emf_2 = 0$$

Using substitution and simplifying, this becomes:

$$6I_1 + 2I_3 - 45 = 0$$

In this case, the signs were reversed compared with the other loop, because elements are traversed in the opposite direction.

We now have three equations that can be used in a system. The second will be used to define I_2, and can be rearranged to:

$$I_2 = 6 - 2I_1$$

The third equation can be used to define I_3, and can be rearranged to:

$$I_3 = 22.5 - 3I_1$$

Substituting the new definitions of I_2 and I_3 (which are both in common terms of I_1), into the first equation ($I_1=I_2+I_3$), we get:

$$I_1 = (6 - 2I_1) + (22.5 - 3I_1) = 28.5 - 5I_1$$

Simplifying, we find that I_1=4.75 A. Inserting this value into the other two equations, we find that I_2=-3.50 A and I_3=8.25 A.

Source Transformation

Source Transformation simply means replacing one source by an equivalent source. A practical voltage source can be transformed into an equivalent practical current source and similarly a practical current source into voltage source.

Any practical voltage source or simply a voltage source consists of an ideal voltage source in series with an internal resistance or impedance (for any ideal source this impedance will be zero), the output voltage becomes independent of the load current. Cells, batteries and generators are the example of the voltage source.

For any practical current source or simply current source, is an ideal current source in parallel with the internal resistance or impedance, for ideal current source this parallel impedance is infinity. The semiconductor devices like transistors, etc are treated as a current source or an output produce by direct or alternating voltage source is called as direct and alternating current source, respectively.

The voltage and current source are mutually transferable or in other words the source transformation i.e. voltage to current source and current to voltage source can be done. Let us understand this by considering a circuit given below:

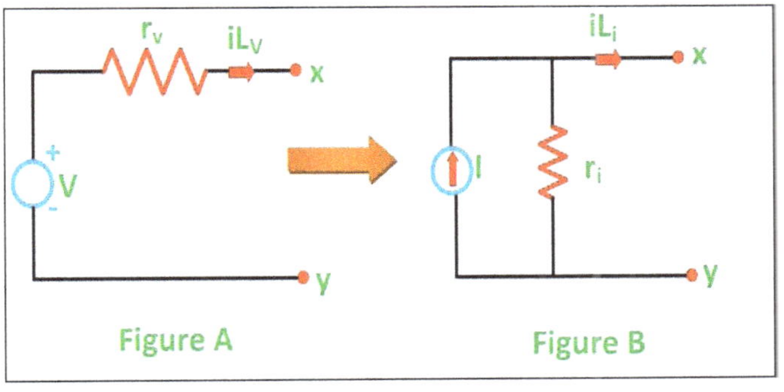

Figure A represents a practical voltage source in series with the internal resistance r_v, while figure B represents a practical current source with parallel internal resistance r_i.

For the practical voltage source the load current will be given by the equation,

$$iL_v = \frac{V}{r_V + r_L}$$

Where,

iL_v is the load current for the practical voltage source.

V is the voltage.

r_v is the internal resistance of the voltage source.

r_L is the load resistance.

It is assumed that the load resistance r_L is connected at the terminal x-y. Similarly for the practical current source the load current is given as:

$$iL_i = I\frac{r_i}{r_i + r_L}$$

Where,

iL_i is the load current for the practical current source.

I is the current.

r_i is the internal resistance of the current source.

r_L is the load resistance connected across the terminal x-y in the figure B.

Two sources become identical, when we will equate the above two equation,

$$\frac{V}{r_V + r_L} = I\frac{r_i}{r_i + r_L}$$

However, for the current source the terminal voltage at x-y would be Iri, x-y terminal are open. i.e.,

$$V = I \times r_i$$

Therefore, we will get,

$$r_V + r_L = r_i + r_L \text{ or}$$
$$r_V = r_L$$

Therefore, for any practical voltage source, if the ideal voltage be V and internal resistance be r_v, the voltage source can be replaced by a current source I with the internal resistance in parallel with the current source.

Source Transformation : Conversion of Voltage Source into Current Source.

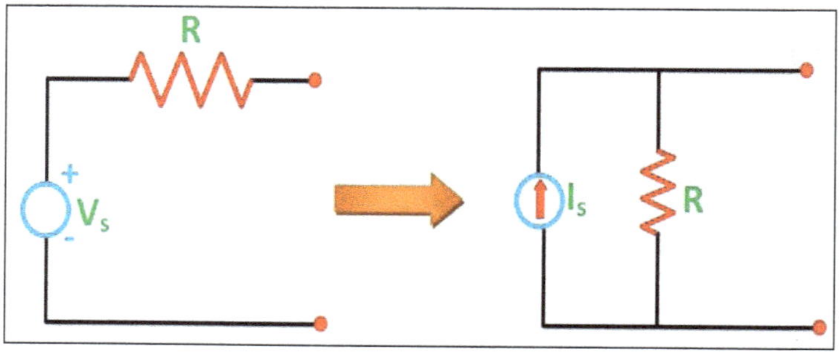

When the voltage source is connected with the resistance in series and it has to be converted into the current source than the resistance is connected in parallel with the current source as shown in the above figure.

Where, $I_s = V_s/R$

Conversion of Current Source into Voltage Source

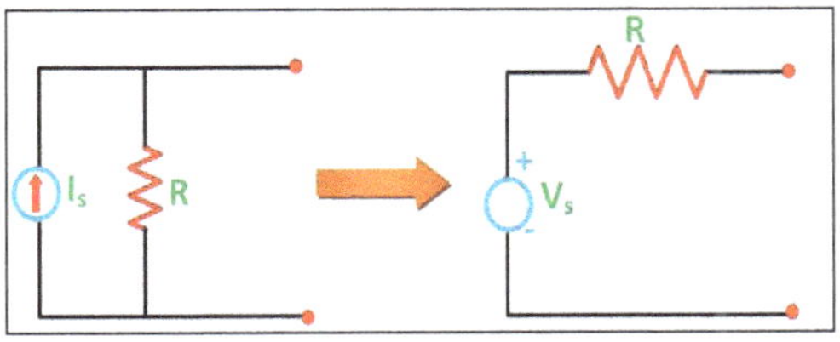

In the above circuit diagram a current source which is connected in parallel with the resistance is transformed into a voltage source by placing the resistance in series with the voltage source.

Where, $V_s = I_s / R$.

Substitution Theorem

Substitution Theorem states that the voltage across any branch or the current through that branch of a network being known, the branch can be replaced by the combination of various elements that will make the same voltage and current through that branch. In other words, the Substitution Theorem says that for branch equivalence, the terminal voltage and current must be same.

The concept of the theorem is based on the substitution of one element from another element. This theorem gives intuition on the behaviour of the circuit. It also helps in proving several other theorems. But the substitution theorem cannot use for solving the theorem which has more than two sources which are neither connected in series nor parallel.

Explanation of Substitution Theorem

Simply one can say that the Substitution Theorem is the replacement of one element with another equivalent element. In a network, if any element is substituted or replaced by a voltage or current source whose voltage and current across or through that element will remain unchanged as the previous network.

Let us understand the theorem with the help of the circuit diagram shown below:

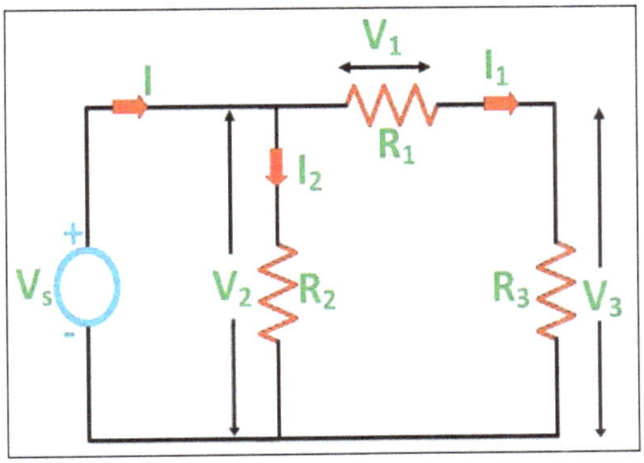

Here various resistances R_1, R_2, R_3 is connected across with the voltage source (V). Current I is flowing through the circuit, which is divided into current I_1 flowing through the resistance R_1 and the current I_2 flowing through the resistance R_2. V_1, V_2 and V_3 are the voltage drop across the resistance R_1, R_2 and R_3 respectively.

Now if the resistance R_3 is substituted by the voltage source V_3 as shown in the circuit diagram below,

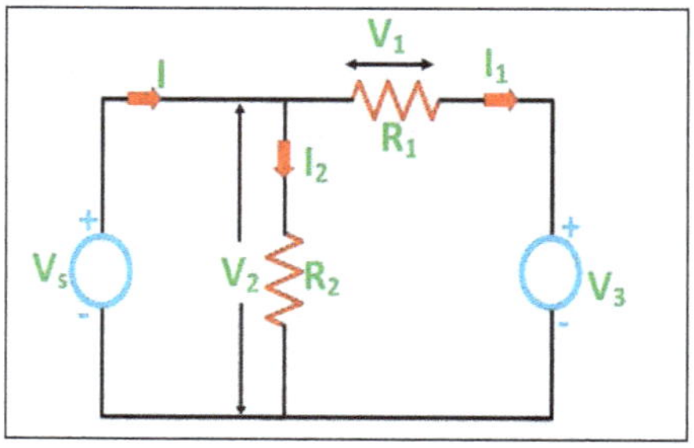

In the circuit diagram shown below the resistance, R_3 is replaced by the current flowing through that element, i.e. I_1.

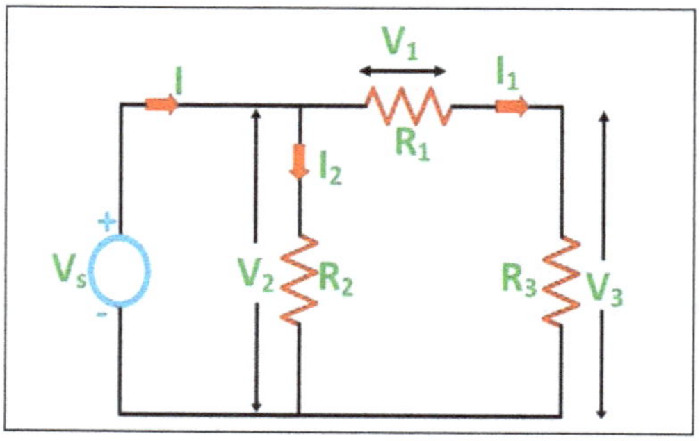

In both the cases shown above if the element is substituted by the voltage source or the current source, then also, the initial conditions of the circuit does not alter. This means that the voltage across the resistance and current flowing through the resistance unaltered even if they are substituted by other sources.

Steps for solving Network using Substitution Theorem

Step 1 – First obtain the concerned branch voltage and current flowing through the branch given by V_{xy} and I_{xy} as shown below in the figure A.

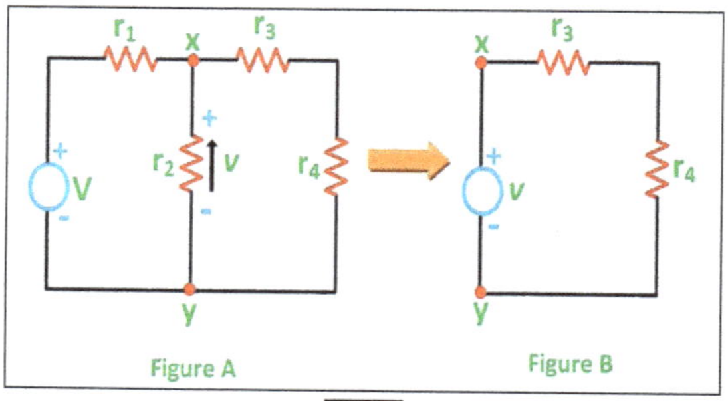

Step 2 – The branch may be substituted by an independent voltage source as shown above in figure B.

Step 3 – Similarly the branch may be replaced by an independent current source as shown below in the circuit diagram C.

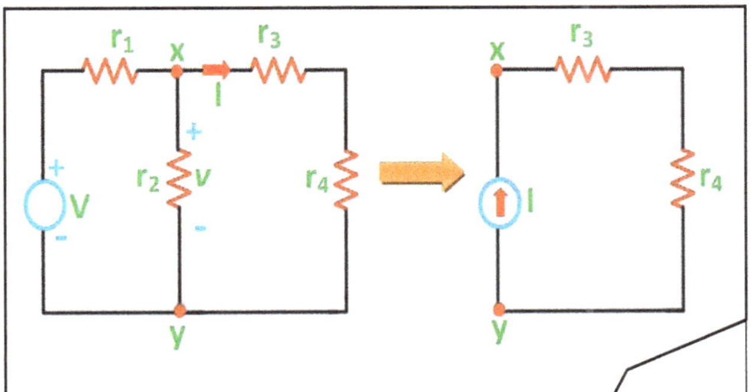

Figure C

Step 4 – Hence it is seen that the voltage drop and the current flowing through the circuit A shown above will be same if it is substituted by any independent voltage or current source shown in the figure B and C.

Norton's Theorem

Norton's theorem states that any linear complex electrical circuit can be reduced into a simple electric circuit with one current and resistance connected in parallel. For understanding in depth regarding norton theory, let us consider Norton's theorem examples as follows:

Examples

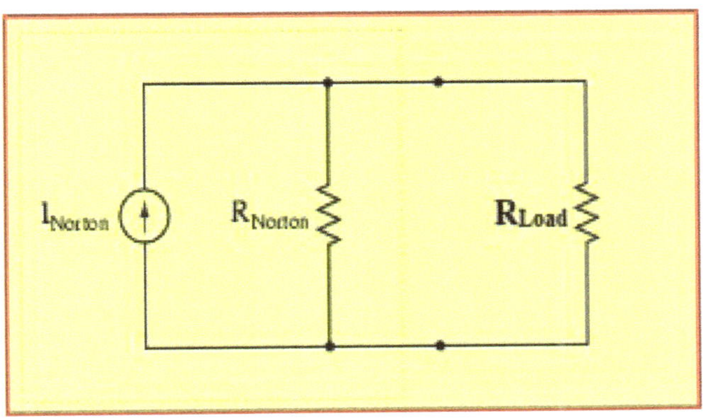

Primarily, let us consider a simple electrical circuit that consists of two voltage sources and three resistors which are connected as shown in the above figure. The above circuit consists of three resistors among which R_2 resistor is considered as load. Then, the circuit can be represented as shown below.

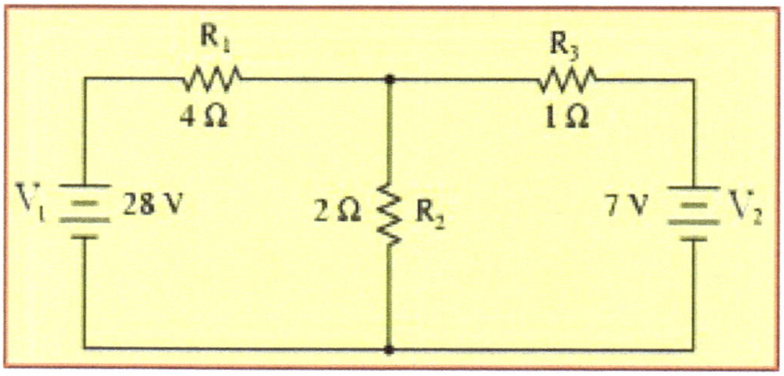

Nortons Theorem Example Circuit with Load Resistor.

We know that, if the load changes, then the calculation of various parameters of electric circuits is difficult. So, network theorems are used for calculating the network parameters easily.

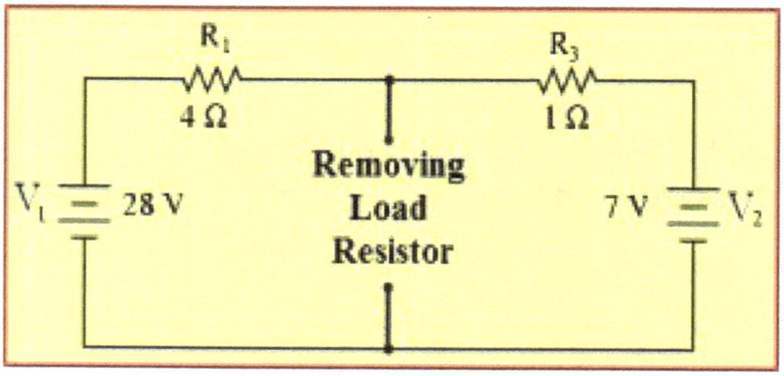

Nortons Theorem Example Circuit After Removing Load Resistor.

In this Norton's theorem also we follow the procedure similar to thevenins theorem (up to some extent). Here, primarily remove the load (consider resistor R_2=2 Ohms as load in the circuit) as shown in the above figure. Then, short circuit the load terminals with a wire (exactly opposite to the procedure that we follow in thevenins theorem, i.e., open circuit of load terminals) as shown in the below figure. Now, calculate the resultant current (current through resistors R_1, R_3, and short circuit line after removing R_2) as shown in the figure below:

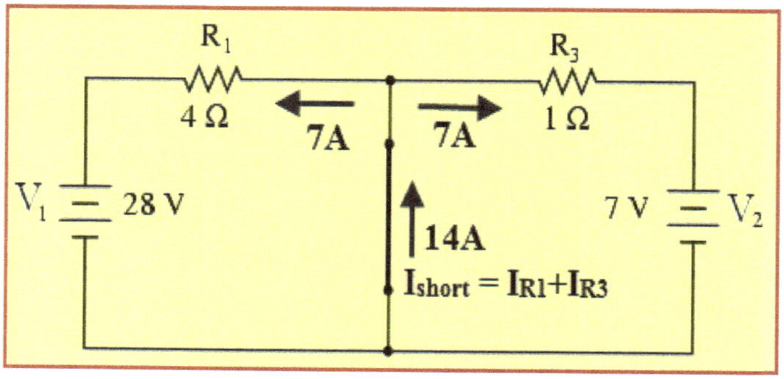

Current Through R1, R3, and Short Circuited Load.

From the above figure, the Nortons source current is equal to 14A which is used in the Norton's equivalent circuit as shown in the below figure. Norton's theorem equivalent circuit consists of the Norton current source (INorton) in parallel with Norton's equivalent resistance (RNorton) and load (here R_2=2Ohms).

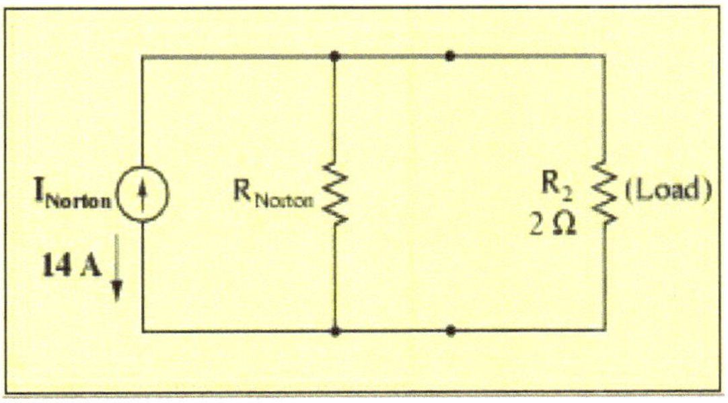

Nortons Equivalent Circuit with INorton, RNorton, RLoad.

This Nortorn's theorem equivalent circuit is a simple parallel circuit as shown in the figure. Now, for calculating Norton's equivalent resistance we have to follow two procedures such as Thevenins theorem and Superposition theorem.

Primarily, remove the load resistance (similar to thevenins theorem step of calculating thevenins resistance). Then, replace voltage sources with short circuit (wires in case of ideal voltage sources and in case of practical voltage sources their internal resistances are used). Similarly, current sources with open circuit (breaks in case of ideal current sources and in case of practical current sources their internal resistances are used). Now, the circuit becomes as shown in the figure below and it is a simple parallel circuit with resistors.

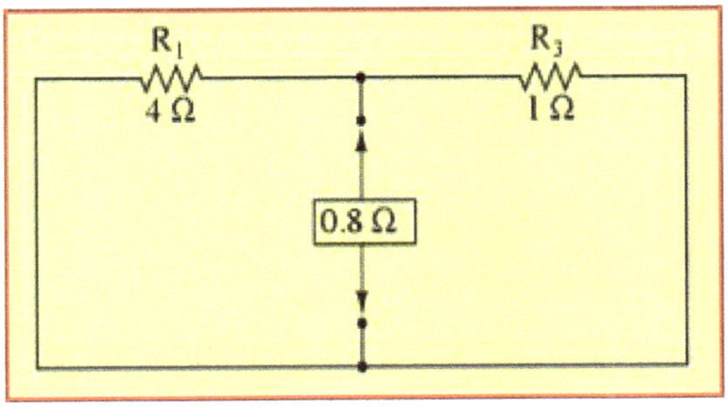

Finding Nortons Resistance.

As the resistors R_1 and R_3 are parallel to each other, the value of Norton's resistance is equal to parallel resistance value of R_1 and R_3. Then, the total Norton's theorem equivalent circuit can be represented as shown in the circuit below:

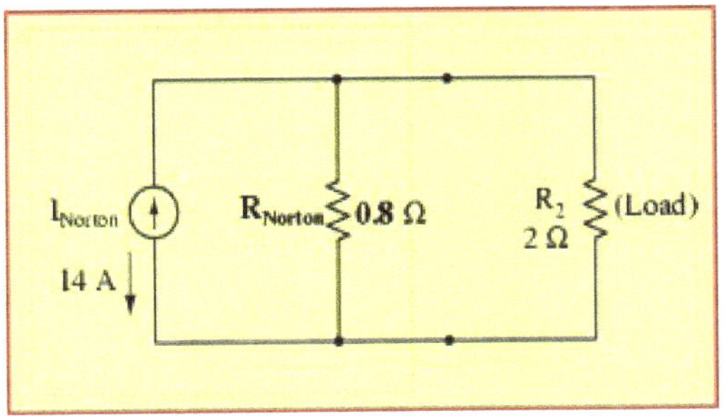

Norton's Theorem Equivalent Circuit.

The formula for calculating the load current, Iload can be calculated using various basic laws such as Ohm's law, Krichhoff's voltage law, and Krichhoff's current law.

Thus, the current passing through the load resistor R load (R_2) is given by,

$$I_L = \frac{I_N . R_N}{R_N + R_L}$$

Load Current formula.

Where,

I_N = Norton's current (14A).

R_N = Norton's resistance (0.8 Ohms).

R_L = Load resistance (2 Ohms).

Therefore, I load = current passing through load resistance = 4A.

Similarly, the large, complex, linear networks with several numbers of sources (current or voltage sources) and resistors can be reduced to simple parallel circuits with single current source in parallel with Norton's resistance and load.

Thus, the Norton's equivalent circuit with Rn and In can be determined and a simple parallel circuit can be formed (from a complex network circuit). The calculations of the circuit parameters can be easily analyzed. If one resistance in the circuit is changed rapidly (load), then the Norton's theorem can be used to perform calculations easily.

Compensation Theorem

Compensation Theorem states that in a linear time invariant network when the resistance (R) of an uncoupled branch, carrying a current (I), is changed by (ΔR). The currents in all the branches would change and can be obtained by assuming that an ideal voltage source of (V_C) has been connected such that V_C= I (ΔR) in series with (R + ΔR) when all other sources in the network are replaced by their internal resistances.

In Compensation Theorem, the source voltage (V_C) opposes the original current. In simple words compensation theorem can be stated as – the resistance of any network can be replaced by a voltage source, having the same voltage as the voltage drop across the resistance which is replaced.

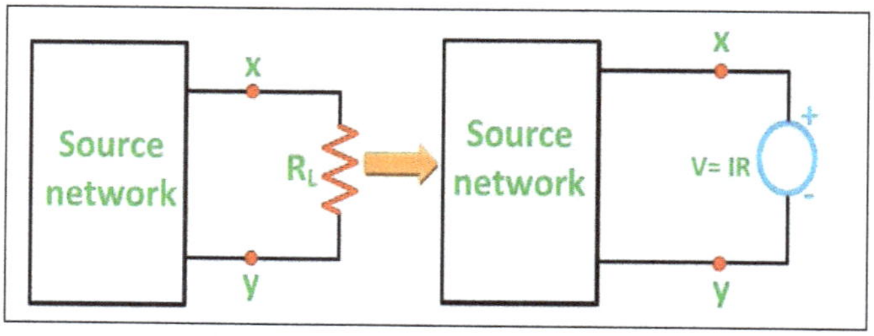

Let us assume a load R_L be connected to a DC source network whose Thevenin's equivalent gives V_o as the Thevenin's voltage and R_{TH} as the Thevenin's resistance as shown in the figure below:

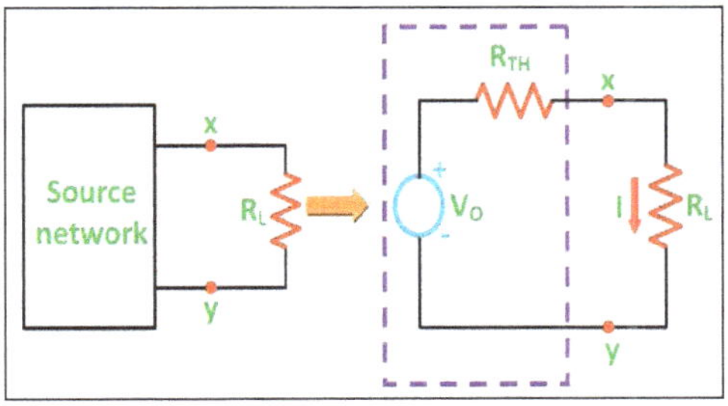

Here,

$$I = \frac{V_0}{R_{TH} + R_L}$$

Let the load resistance R_L be changed to $(R_L + \Delta R_L)$. Since the rest of the circuit remains unchanged, the Thevenin's equivalent network remains the same as shown in the circuit diagram below:

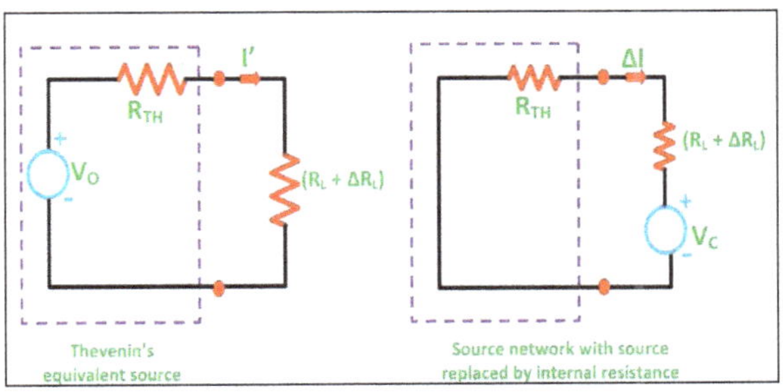

Here,

$$I' = \frac{V_0}{R_{TH} + (R_L + \Delta R_L)}$$

The change of current being termed as ΔI,

Therefore,

$$\Delta I = I' - I$$

Putting the value of I' and I from the above three equation we will get the following equation,

$$\Delta I = \frac{V_0}{R_{TH} + (R_L + \Delta R_L)} - \frac{V_0}{R_{TH} + R_L}$$

$$\Delta I = \frac{V_0\{R_{TH} + R_L - (R_{TH} + R_L + \Delta R_L)\}}{(R_{TH} + R_L + \Delta R_L)(R_{TH} + R_L)}$$

$$\Delta I = -\left[\frac{V_0}{R_{TH} + R_L}\right]\frac{\Delta R_L}{R_{TH} + R_L + \Delta R_L}$$

Now, putting the value of I from the equation mentioned in the starting of this topic in the equation above, we will get the following equation,

$$\Delta I = -\frac{I \Delta R_L}{R_{TH} + R_L + \Delta R_L}$$

As we know, $V_C = I \Delta R_L$ and is known as compensating voltage.

Therefore, the above equation becomes,

$$\Delta I = -\frac{-V_C}{R_{TH} + R_L + \Delta R_L}$$

Hence, Compensation Theorem tells that with the change of branch resistance, branch currents changes and the change is equivalent to an ideal compensating voltage source in series with the branch opposing the original current, all other sources in the network being replaced by their internal resistances.

Superposition Theorem

Superposition theorem is based on the concept of linearity between the response and excitation of an electrical circuit. It states that the response in a particular branch of a linear circuit when multiple independent sources are acting at the same time is equivalent to the sum of the responses due to each independent source acting at a time.

In this method, we will consider only one independent source at a time. So, we have to eliminate the remaining independent sources from the circuit. We can eliminate the voltage sources by shorting their two terminals and similarly, the current sources by opening their two terminals.

Therefore, we need to find the response in a particular branch 'n' times if there are 'n' independent sources. The response in a particular branch could be either current flowing through that branch or voltage across that branch.

Procedure of Superposition Theorem

Follow these steps in order to find the response in a particular branch using superposition theorem:

- Step 1: Find the response in a particular branch by considering one independent source and eliminating the remaining independent sources present in the network.
- Step 2: Repeat Step 1 for all independent sources present in the network.
- Step 3: Add all the responses in order to get the overall response in a particular branch when all independent sources are present in the network.

Example: Find the current flowing through 20 Ω resistor of the following circuit using superposition theorem.

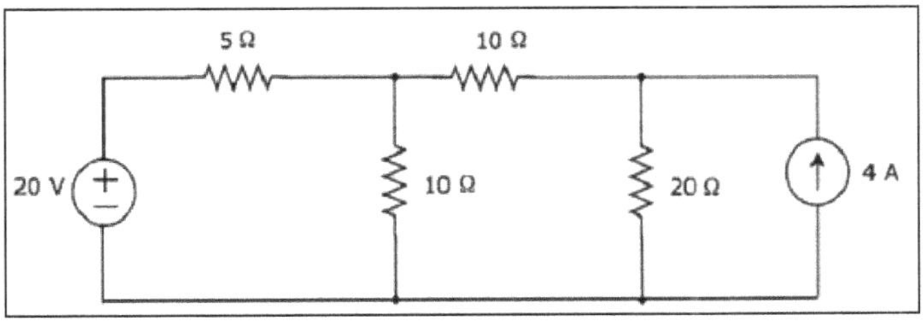

Step 1

Let us find the current flowing through 20 Ω resistor by considering only 20 V voltage source. In this case, we can eliminate the 4 A current source by making open circuit of it. The modified circuit diagram is shown in the following figure.

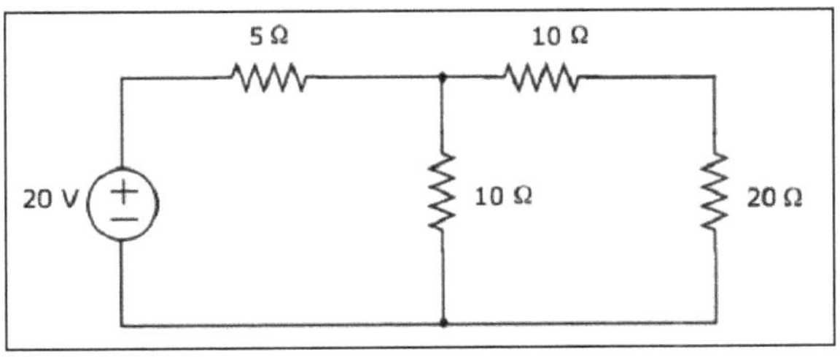

There is only one principal node except Ground in the above circuit. So, we can use nodal analysis method. The node voltage V_1 is labelled in the following figure. Here, V_1 is the voltage from node 1 with respect to ground.

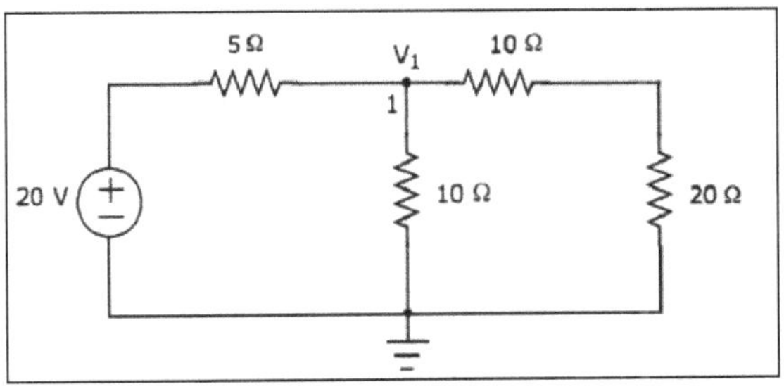

The nodal equation at node 1 is,

$$\frac{V_1 - 20}{5} + \frac{V_1}{10} + \frac{V_1}{10+20} = 0$$

$$\Rightarrow \frac{6V_1 - 120 + 3V_1 + V_1}{30} = 0$$

$$\Rightarrow 10V_1 = 120$$

$$\Rightarrow V_1 = 12V$$

The current flowing through 20 Ω resistor can be found by doing the following simplification.

$$I_1 = \frac{V_1}{10+20}$$

Substitute the value of V_1 in the above equation.

$$I_1 = \frac{12}{10+20} = \frac{12}{30} = 0.4\,A$$

Therefore, the current flowing through 20 Ω resistor is 0.4 A, when only 20 V voltage source is considered.

Step 2

Let us find the current flowing through 20 Ω resistor by considering only 4 A current source. In this case, we can eliminate the 20 V voltage source by making short-circuit of it. The modified circuit diagram is shown in the following figure.

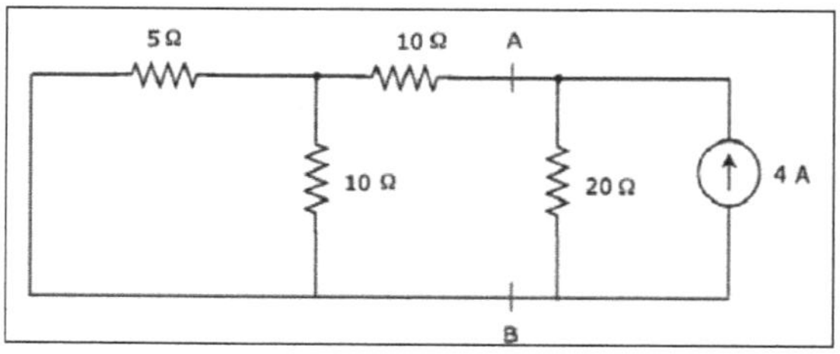

$$R_{AB} = \left(\frac{5\times 10}{5+10}\right) + 10 = \frac{10}{3} + 10 = \frac{40}{3}\Omega$$

The simplified circuit diagram is shown in the following figure.

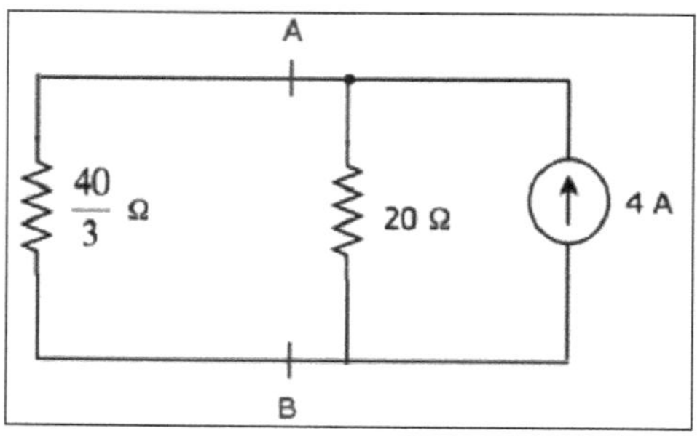

We can find the current flowing through 20 Ω resistor, by using current division principle.

$$I_2 = I_S\left(\frac{R_1}{R_1+R_2}\right)$$

Substitute I_S=4A, $R_1 = \frac{40}{3}$ Ω and R_2=20Ω in the above equation.

$$I_2 = 4\left(\frac{\frac{40}{3}}{\frac{40}{3}+20}\right) = 4\left(\frac{40}{100}\right) = 1.6A$$

Therefore, the current flowing through 20 Ω resistor is 1.6 A, when only 4 A current source is considered.

Step 3

We will get the current flowing through 20 Ω resistor of the given circuit by doing the addition of two currents that we got in step 1 and step 2. Mathematically, it can be written as:

$$I = I_1 + I_2$$

Substitute, the values of I_1 and I_2 in the above equation.

$$I = 0.4 + 1.6 = 2A$$

Therefore, the current flowing through 20 Ω resistor of given circuit is 2A.

Thevenin's Theorem

Thevenin's Theorem states that – any complicated network across its load terminals can be substituted by a voltage source with one resistance in series. This theorem helps in the study of the variation of current in a particular branch when the resistance of the branch is varied while the remaining network remains the same for example designing of electronics circuits.

A more general statement of Thevenin's Theorem is that any linear active network consisting of independent or dependent voltage and current source and the network elements can be replaced by an equivalent circuit having a voltage source in series with a resistance, that voltage source being the open circuited voltage across the open circuited load terminals and the resistance being the internal resistance of the source.

In other words, the current flowing through a resistor connected across any two terminals of a network by an equivalent circuit having a voltage source E_{th} in series with a resistor R_{th}. Where Eth is the open circuit voltage between the required two terminals called the Thevenin voltage and the R_{th} is the equivalent resistance of the network as seen from the two terminal with all other sources replaced by their internal resistances called Thevenin resistance.

Explanation of Thevenin's Theorem

The Thevenin's statement is explained with the help of a circuit shown below.

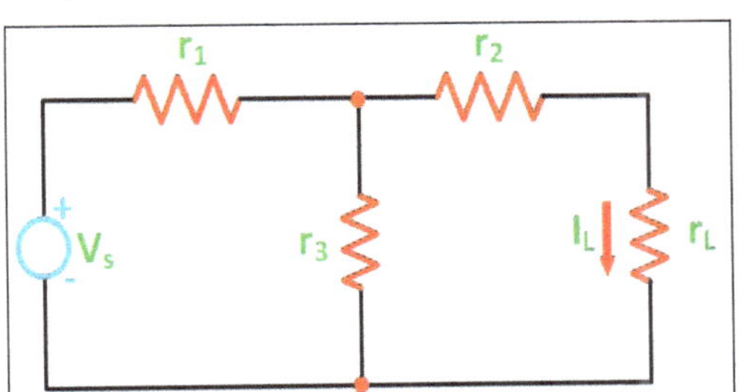

Let us consider a simple DC circuit as shown in the figure above, where we have to find the load current I_L by the Thevenin's theorem. In order to find the equivalent voltage source, r_L is removed from the circuit as shown in the figure below and V_{oc} or V_{TH} is calculated.

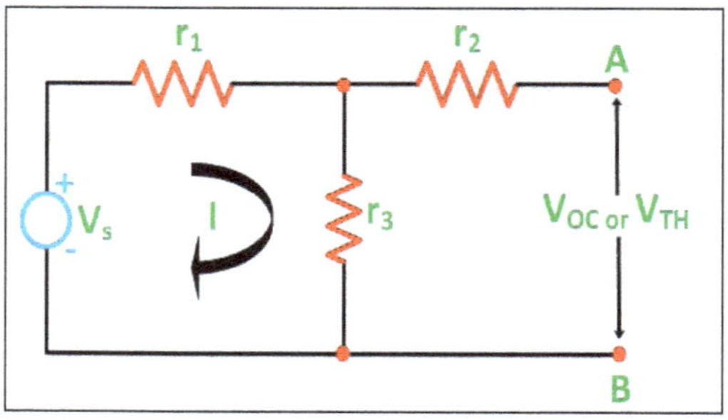

$$V_{OC} = I\, r_3 = \frac{V_S}{r_1 + r_3} r_3$$

Now, to find the internal resistance of the network (Thevenin's resistance or equivalent resistance) in series with the open circuit voltage V_{OC}, also known as Thevenin's voltage V_{TH}, the voltage source is removed or we can say it is deactivated by a short circuit (as the source does not have any internal resistance) as shown in the figure below:

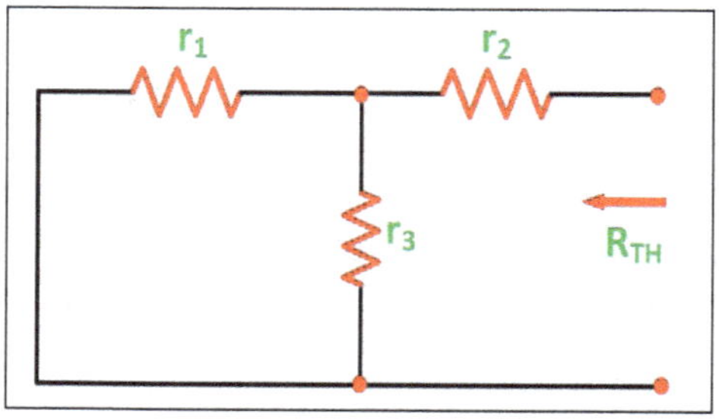

$$R_{TH} = r_2 + \frac{r_1 r_3}{r_1 + r_3}$$

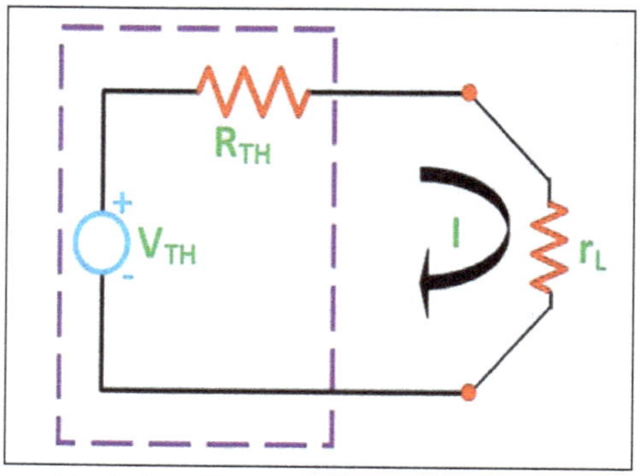

Equivalent Circuit of Thevenin's Theorem.

As per Thevenin's Statement, the load current is determined by the circuit shown above and the equivalent Thevenin's circuit is obtained.

The Load current I_L is given as,

$$I_L = \frac{V_{TH}}{R_{TH} + r_L}$$

Where,

V_{TH} is the Thevenin's equivalent voltage. It is an open circuit voltage across the terminal AB known as load terminal R_{TH} is the Thevenin's equivalent resistance, as seen from the load terminals where all the sources are replaced by their internal impedance r_L is the load resistance.

Steps for Solving Thevenin's Theorem

- Step 1: First of all remove the load resistance r_L of the given circuit.
- Step 2: Replace all the impedance source by their internal resistance.
- Step 3: If sources are ideal then short circuit the voltage source and open the current source.
- Step 4: Now find the equivalent resistance at the load terminals know as Thevenin's Resistance (R_{TH}).
- Step 5: Draw the Thevenin's equivalent circuit by connecting the load resistance and after that determine the desired response.

This theorem is possibly the most extensively used networks theorem. It is applicable where it is desired to determine the current through or voltage across any one element in a network. The Thevenin's Theorem is an easy way to solve the complicated network.

Example

For example, consider the following circuit;

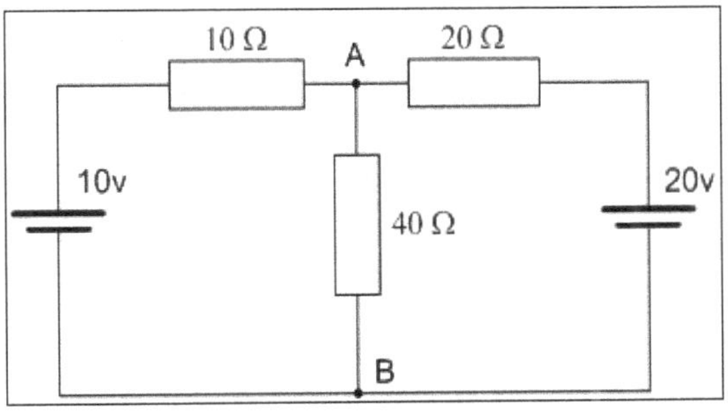

Firstly, we have to remove the centre 40Ω resistor and short out (not physically as this would be dangerous) all the emf's connected to the circuit, or open circuit any current sources. The value of resistor Rs is found by calculating the total resistance at the terminals A and B with all the emf's removed, and the value of the voltage required Vs is the total voltage across terminals A and B with an open circuit and no load resistor Rs connected. Then, we get the following circuit.

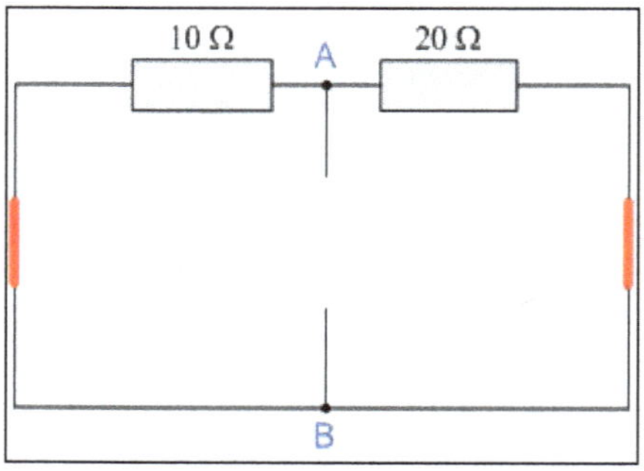

Find the Equivalent Resistance (Rs):

10ΩResistor in parallel with the 20ΩResistor,

$$R_T = \frac{R_1 \times R_2}{R_1 + R_2} = \frac{20 \times 10}{20 + 10} = 6.67\Omega$$

Find the Equivalent voltage (Vs):

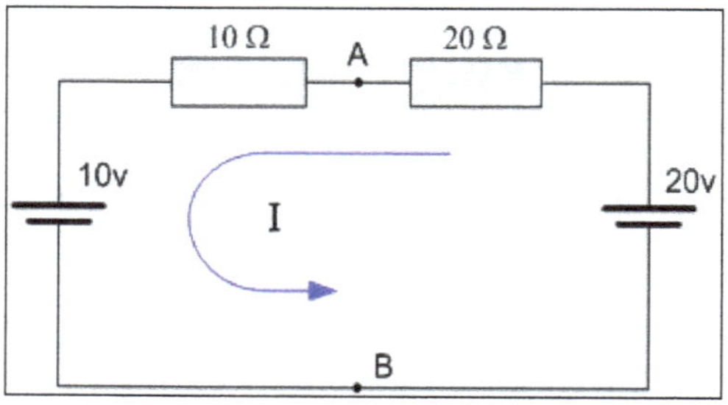

We now need to reconnect the two voltages back into the circuit, and as $V_S = V_{AB}$ the current flowing around the loop is calculated as:

$$I = \frac{20v - 10v}{20\Omega + 10\Omega} = 0.33amps$$

So the voltage drop across the 20Ω resistor can be calculated as:

V_{AB} = 20 - (20Ω x 0.33amps) = 13.33 volts.

Then the Thevenin's Equivalent circuit is shown below with the 40Ω resistor connected.

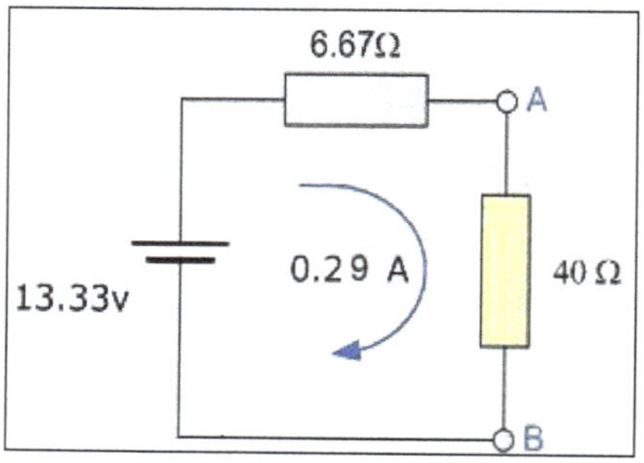

And from this the current flowing in the circuit is given as:

$$I = \frac{13.33v}{6.67\Omega + 40\Omega} = 0.29amps$$

Thevenin's theorem can be used as a circuit analysis method and is particularly useful if the load is to take a series of different values. It is not as powerful as Mesh or Nodal analysis in larger networks because the use of Mesh or Nodal analysis is usually necessary in any Thevenin exercise, so it might well be used from the start. However, Thevenin's equivalent circuits of Transistors, Voltage Sources such as batteries etc, are very useful in circuit design.

Applications of Thevenin's Theorem

- Thevenin's Theorem is especially useful in analyzing power systems and other circuits where one particular resistor in the circuit (called the "load" resistor) is subject to change, and re-calculation of the circuit is necessary with each trial value of load resistance, to determine voltage across it and current through it.
- Source modeling and resistance measurement using the Wheatstone bridge provide applications for Thevenin's theorem.

Practical Limitations

- Many, if not most circuits are only linear over a certain range of values, thus the Thévenin equivalent is valid only within this linear range and may not be valid outside the range.
- The Thévenin equivalent has an equivalent I-V characteristic only from the point of view of the load.
- The power dissipation of the Thévenin equivalent is not necessarily identical to the power dissipation of the real system. However, the power dissipated by an external resistor between the two output terminals is the same however the internal circuit is represented.

References

- Network-analysis-for-electric-circuits: electricalengineeringforbeginners.blogspot.com, Retrieved 08 March 2020
- Electric-circuit-or-electrical-network: electrical4u.com, Retrieved 23 July 2020
- What-is-substitution-theorem: circuitglobe.com, Retrieved 09 July 2020
- Active-and-passive-elements-of-electrical-circuit: electrical4u.com, Retrieved 18 April 2020
- Electrical-impedance-and-its-applications: elprocus.com, Retrieved 09 July 2020
- What-is-an-electrical-circuit, electricity, science: eschooltoday.com, Retrieved 28 May 2020

4

Common Electric Machines

An electric machine is a general term for machines using electromagnetic forces to convert electricity to mechanical power and vice versa. A few electric machines closely examined and explained in this chapter are the rotary converter, metadyne, Dc motor, Ac motor, induction motor, transformer and doubly-fed electric machine.

Rotary Converter

A rotary converter is a type of electrical machine which acts as a mechanical rectifier, inverter or frequency converter.

Rotary converters were used to convert alternating current (AC) to direct current (DC), or DC to AC power, before the advent of chemical or solid state power rectification and inverting. They were commonly used to provide DC power for commercial, industrial and railway electrification from an AC power source.

Rotary converters are still used as frequency converters.

Principles of Operation

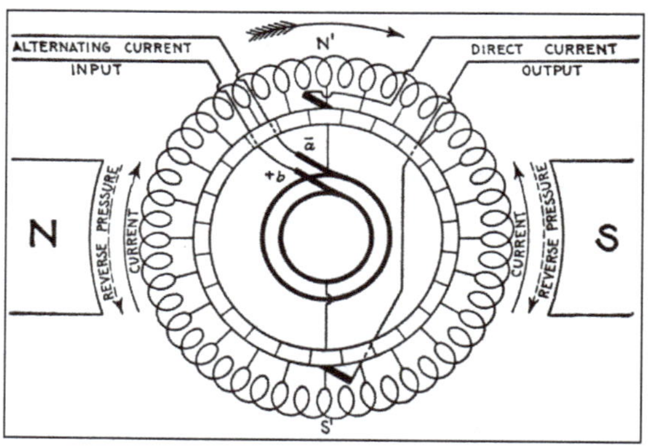

Wiring schematic for a simplified bipolar field Gramme ring single-phase to direct current rotary converter. (In actual use, the converter is drum-wound and uses a multipolar field).

The rotary converter can be thought of as a motor-generator, where the two machines share a single rotating armature and set of field coils. The basic construction of the rotary converter consists of a DC generator (dynamo) with a set of slip rings tapped into its rotor windings at evenly spaced intervals. When a dynamo is spun the electric currents in its rotor windings alternate as it rotates in the magnetic field of the stationary field windings. This alternating current is rectified by means of a commutator which allows DC current to be extracted from the rotor. This principle is taken advantage of by energizing the same rotor windings with AC power which causes the machine to act as a synchronous AC motor. The rotation of the energized coils excites the stationary field windings producing part of the DC current. The other part is AC current from the slip rings which is directly rectified into DC by the commutator. This makes the rotary converter a hybrid dynamo and mechanical rectifier. When used in this way it is referred to as a synchronous rotary converter or simply a synchronous converter. The AC slip rings also allow the machine to act as an alternator.

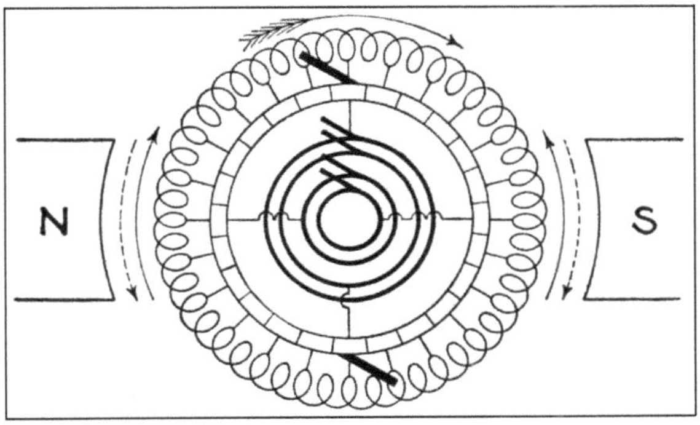

Wiring schematic for a simpllfied two-phase to direct current rotary converter, with the second phase connected at right angles to the first.

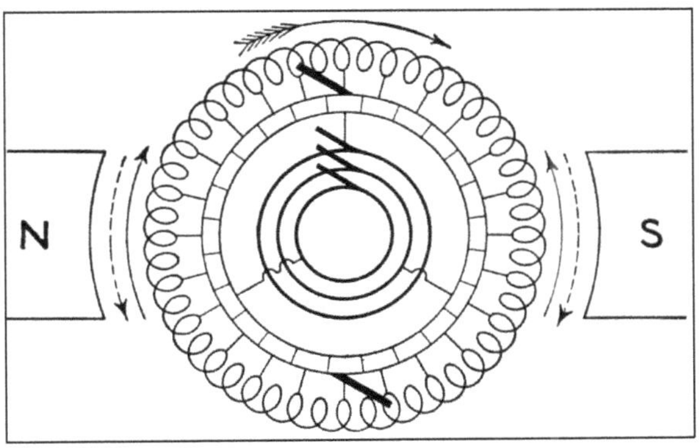

Wiring schematic for a simpllfied three-phase to direct current rotary converter, with the phases separated by 120 degrees on the commutator.

The device can be reversed and DC applied to the field and commutator windings to spin the machine and produce AC power. When operated as a DC to AC machine it is referred to as an inverted rotary converter.

One way to envision what is happening in an AC-to-DC rotary converter is to imagine a rotary reversing switch that is being driven at a speed that is synchronous with the power line. Such a switch could rectify the AC input waveform with no magnetic components at all save those driving the switch. The rotary converter is somewhat more complex than this trivial case because it delivers near-DC rather than the pulsating DC that would result from just the reversing switch, but the analogy may be helpful in understanding how the rotary converter avoids transforming all of the energy from electrical to mechanical and back to electrical.

The advantage of the rotary converter over the discrete motor-generator set is that the rotary converter avoids converting all of the power flow into mechanical energy and then back into electrical energy; some of the electrical energy instead flows directly from input to output, allowing the rotary converter to be much smaller and lighter than a motor-generator set of an equivalent power-handling capability. The advantages of a motor-generator set include adjustable voltage regulation which can compensate for voltage drop in the supply network; it also provided complete power isolation, harmonics isolation, greater surge and transient protection, and sag (brownout) protection through increased momentum.

In this first illustration of a single-phase to direct-current rotary converter, it may be used five different ways:

- If the coil is rotated, alternating currents can be taken from the collector rings, and it is called an alternator.
- If the coil is rotated, direct current can be taken from the commutator, and it is called a dynamo.
- If the coil is rotated, two separate currents can be taken from the armature, one providing direct current and the other providing alternating current. Such a machine is called a *double current generator*.
- If a direct current is applied to the commutator, the coil will begin to rotate as a commutated electric motor and an alternating current can be taken out of the collector rings. This is called an *inverted rotary converter*.
- If the machine is brought up to synchronous speed by external means and if the direction of the current through the armature has the correct relationship to the field coils, then the coil will continue to rotate in sychronism with the alternating current as a synchronous motor. A direct current can be taken from the commutator. When used this way, it is called a *rotary converter*.

Applications

A typical use for an AC/DC converter was for railway electrification, where utility power was supplied as alternating current but the trains were designed to work on direct current. Before the invention of mercury arc rectifiers and high-power semiconductor rectifiers, this conversion could only be accomplished using motor-generators or rotary converters.

Railroad Rotary Converter from Illinois Railway Museum.

Around 1900 most machinery and appliances were operated by DC power which was provided by rotary converter substations for residential, commercial and industrial consumption. Rotary converters provided high current DC power for industrial electrochemical processes such as electroplating. Steel mills needed large amounts of on-site DC power for their main roll drive motors. Similarly, paper mills and printing presses required direct current to start and stop their motors in perfect synchronization to prevent tearing the sheet.

Obsolescence

AC to DC synchronous rotary converters were made obsolete by mercury arc rectifiers in the 1930s and later on by semiconductor rectifiers in the 1960s. Some of the original New York City Subway substations using synchronous rotary converters operated until 1999. Compared to the rotary converter, the mercury arc and semiconductor rectifiers did not need daily maintenance, manual synchronizing for parallel operation, nor skilled personnel, and they provided clean DC power. This enabled the new substations to be unmanned, only requiring periodic visits from a technician for inspection and maintenance.

AC replaced DC in most applications and eventually the need for local DC substations diminished along with the need for rotary converters. Many DC customers converted to AC power, and on-site solid-state DC rectifiers were used to power the remaining DC equipment from the AC supply.

Self-balancing Dynamo

The self-balancing dynamo is of similar construction to the single- and two-phase rotary converter. It was commonly used to create a completely balanced three-wire 120/240-volt DC electrical supply. The AC extracted from the slip rings was fed into a transformer with a single center-tapped winding. The center-tapped winding forms the DC neutral wire. It needed to be driven by a mechanical power source, such as a steam engine, diesel engine, or electric motor. It could be considered a rotary converter used as a double current generator; the AC current was used to balance the DC neutral wire.

Metadyne

A metadyne is a direct current electrical machine with two pairs of brushes. It can be used as an amplifier or rotary transformer. It is similar to a third brush dynamo but has additional regulator or "variator" windings. It is also similar to an amplidyne except that the latter has a compensating winding which fully counteracts the effect of the flux produced by the load current. The technical description is "a cross-field direct current machine designed to utilize armature reaction". A metadyne can convert a constant-voltage input into a constant-current, variable-voltage output.

History

The word *metadyne* is derived from the Greek words for conversion of power. While the name is believed to have been coined by Joseph Maximus Pestarini in a paper which he submitted to the Montefiore International Contest at Liège, Belgium in 1928, the type of machine which it described had been known since the 1880s. The first known British patent for a direct-current, cross-field generator was obtained by A. I. Gravier of Paris in 1882, and two further patents were obtained by E. Rosenberg in 1904 and 1907. Rosenberg later became the chief electrical engineer for Metropolitan-Vickers, and his machine produced a cross field by applying a short-circuit to an additional set of brushes. M. Osnos looked at the practical arrangements for several such machines in 1907, and in the same year, Felton and Guilleaume obtained a British patent, number 26,607, which described auxiliary windings, armature windings and multiple commutators, although all in fairly general terms. He also indicated that they could be used to transform a constant voltage into a constant current. Other patents were obtained prior to 1910 by Mather & Platt, Brown Boverei and Bruce Peebles.

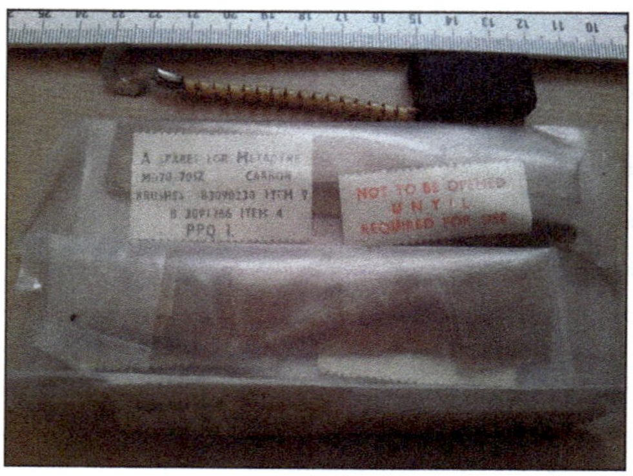

Spare Metadyne carbon brushes. They are individually wrapped and, together with labels, sealed in plastic. Total length including the connecting wire and ring is 115 mm. The thickness of the carbon slab is 8 mm.

Pestarini worked on developing the theory of such machines between 1922 and 1930, although he concentrated on their static characteristics, rather than their dynamic characteristics. He contributed three papers on the subject to *Revue Générale de l'Electricité* in 1930, which included some practical applications. The main one was the use of the constant-current output for the control of traction motors on electric vehicles and the operation of cranes, areas in which he had some practical experience, following trials in conjunction with the Alsthom Company in France. In 1930, he made a trip to Britain, and the Metropolitan-Vickers company took his ideas and developed a working system. Unlike Rosenberg's solution, Pestarini, who later became a Professor at the Institute Electrotechnico Nazionale Galileo Ferraris in Turin, connected the additional brushes to an external supply to produce a transformer metadyne. The machine worked as a voltage-to-current amplifier because the flux generated by the current to the load opposed the flux in the control circuit. Development work at Metropolitan-Vickers in the 1930s was led by Arnold Tustin, and the company held the British patents for the Metadyne.

Pestarini also visited the United States in 1930, although there is no record of the system being used there. The General Electric engineers, led by Ernst Alexanderson, were interested, but modified the design by the addition of a compensating winding, which counteracted the effect of the flux produced by the load current. This turned the machine from a voltage-to-current amplifier into a voltage-to-voltage amplifier, and they called the new variant an Amplidyne. The development costs were largely funded by US naval contracts for the development of vertical stabilisers, which were used to improve the aiming and firing of guns on ships. During the same period, the Macfarlane Engineering Company, who were based in Glasgow, developed a variant of the cross field machine quite independently, which they named the *Magnicon*.

Pestarini filed a patent on the metadyne machine in France on 14 January 1932, and submitted it to the United States Patent Office at the end of the year, on 23 December. The US patent was granted on 30 January 1934. He submitted a second US patent for an improved machine in November 1946, which was granted on 10 June 1952.

Operation

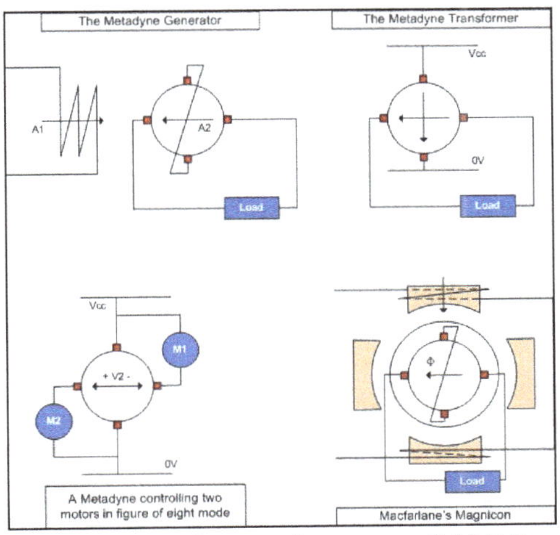

Three arrangements of a metadyne cross-field DC machine, and the construction of Macfarlane's Magnicon.

The diagram shows three arrangements of a metadyne machine. In all cases, compensation windings have been omitted for clarity. The first arrangement represents a one-cycle cross-field machine. In a normal DC machine, the effect of the excitation current generates a flux (A1), which in turn generates a quadrature flux which is at right angles to the exciting flux. By wiring the quadrature brushes together, current is produced in the armature, and the flux that this produces (A2) is again at right angles to the quadrature axis, resulting in an armature reaction which is directly opposed to the original excitation. This feature is fundamental to the machine, and does not depend on its direction of rotation. When the armature reaction is partially compensated by a compensation winding, the uncompensated portion of the armature reaction acts in this way. As the output current rises, it suppresses the effect of the excitation, until it reaches a state where there is just enough excitation to maintain the current. Any further increase would eliminate the flux which sustains its operation, and the current is maintained irrespective of the resistance of the load or the back emf produced by it. The machine thus acts as a constant-current generator, where the current is proportional to the excitation.

The second diagram shows a machine with no excitation winding, but instead, a constant voltage is connected to the quadrature brushes. This produces a flux similar to the one produced by the rotation of the armature in the excitation flux in the first example. The operation of the machine is therefore very similar, with the output current rising until the flux it produces almost counteracts the flux generated by the applied voltage. Tustin has shown that the input and output power are the same, and so the machine transforms the constant-voltage input into a constant-current output. As with the metadyne generator, the Metadyne transformer can be partially compensated, and will continue to operate as a constant-current device until the compensation exceeds 97 per cent.

The third diagram shows a metadyne connected to two separate motors, and this arrangement was often used for the control of traction motors on electric trains. Connecting them in this way reduces the effective loading on the Metadyne, and enables a smaller machine to be installed. The Metadyne acts as a "positive or negative booster". If Vcc is the supply voltage, and V2 is the output voltage of the Metadyne, then the total voltage across the load can vary from 0 to 2·Vcc, as V2 varies between –Vcc and +Vcc. Although the system is prone to the currents in the two halves of the load becoming unbalanced, this can be corrected by the provision of extra series windings, which act like an additional circuit resistance.

Rosenberg Generator

The Rosenberg generator is very similar to the Metadyne generator, both in its construction and its electrical connection. It generally does not have a compensation winding, so that the whole of the armature reaction opposes the initial excitation. Parts of the magnetic circuit are normally not laminated, which creates delays between excitations and fluxes, but the machines are used in applications where a quick response is not essential. Their predominant use has been in trains, where they are axle-driven, and used to provide lighting and the charging of batteries. An axle-driven generator is subject to variable speeds and changes in the direction of rotation, but the characteristics of the machine allow it to produce useful energy down to very low speeds. At slow speeds, the output voltage increases with the square of the speed, but the magnetic circuit soon becomes saturated, resulting in much smaller increases as the speed increases. When used in circuits which include batteries charged from the output, a rectifier or reverse-current cut-out is normally required to prevent discharge of the batteries through the generator at very low speeds, or when the train stops.

Magnicon

The Magnicon, developed by Macfarlane's in Scotland, is similar to the Metadyne, but whereas the latter has a two-pole armature winding, the Magnicon has a four-pole lap winding, and is sometimes referred to a Metadyne with a short-pitched armature winding. They have been supplied to operate hoists and winches on ships. The stator of a Magnicon has four polar projections, spaced at 90 degrees, and one pair of them is excited. The pair of brushes which are on the same axis as the excited poles are short-circuited, resulting in a large current. The magnetomotive force (MMF) of this current acts on the non-excited poles, creating a working flux (Φ) and the output voltage. As with a full-pitch Metadyne, the armature reaction of the output current is 90 degrees out of phase, and therefore opposes the original excitation. Advantages over the normal Metadyne are that the number of exciting and compensating coils is halved to two each per cycle, and the shorter pitch of the coils results in less overhang at the ends of the windings. However, the design creates idle currents in the armature, which result in losses, and on larger machines, where interpoles are required, each interpole must be fitted with two coils, one for each of the brush circuits. Tustin argues that there is little advantage of the Magnicon over the Metadyne for smaller machines, and for larger machines, which require interpoles to be fitted, insufficient analysis has been carried out to make a judgement.

Uses

Metadynes have been used to control the aiming of large guns and for speed control in electric trains, in particular the London Underground O and P Stock. They have been superseded by solid state devices.

Traction Control

In the early 1930s, the London Underground were aware of the development of the metadyne equipment taking place at Metropolitan-Vickers, and the potential for regenerative braking which it provided. Before committing to an untried system, they therefore built a test train, by converting six cars originally built between 1904 and

1907 for the Metropolitan Railway. The work was carried out at Acton Works in 1934. Since a single metadyne unit could be used to control four motors, and each motor car had two motors, they were formed into two-car units, with a driving cab at the outer ends. By coupling the units together, trials of a two-car, four-car and six-car train could be carried out. The metadyne unit weighed about 3 tons, and consisted of three rotating machines, an exciter, a regulator and the actual metadyne machine, which were linked together mechanically. Electrically, the traction supply was fed into the machine, and the output fed the motors, without the need for starting resistances.

A London Underground CP train (in red) at Upminster. They were originally fitted with Metadyne controls and were the first electric multiple units to use regenerative braking.

The test train ran through much of 1935 and 1936, and was tried on nearly all of the electrified tracks on the Metropolitan line and the District line. Once the concept had been proved to be reliable, the train was also used in passenger service. Besides the regenerative braking, the acceleration was found to be particularly smooth. When the decision was taken to proceed with the new system on the O and P stock, the test train was dismantled, and the equipment was fitted to three battery locomotives built by the Gloucester Railway Carriage and Wagon Company, which were part of a batch of nine vehicles supplied between 1936 and 1938. The equipment was particularly suitable for battery locomotives, as the lack of starting resistances reduced the amount of power wasted when starting and stopping frequently. At slow speeds, conventional control systems would often overheat, but the metadyne-equipped locomotives could pull trains weighing 100 tons for long distances at speeds as low as 3 mph (4.8 km/h) without problems. However, the complexity of the equipment, and the difficulty of maintaining the metadyne machine, resulted in the locomotives not being used sufficiently, and they were withdrawn for scrapping in 1977.

The main production run of O Stock consisted of 116 motor cars, which were formed into 58 two-car units. Trials began with a four-car formation on the District line between High Street Kensington and Putney Bridge in September 1937, and a six-car formation began working on the Hammersmith line in January 1938. There were some technical problems, caused by the demands put on the power supply system when a train of six motor cars started, and the amount of power that such a train attempted to return to the system when the regenerative brakes were used. This was partially mitigated by ordering a further 58 trailer cars, and converting each two-car unit into a three-car unit, by inserting a trailer car into the formation. A batch of P Stock was then ordered, to replace the trains on the Metropolitan line. Although O and P Stock units could be coupled together, the metadyne units in particular were not the same, and could not be interchanged between the builds. By the early 1950s, this was a serious disadvantage, when a series of failures occurred, which required extensive repairs. A decision was taken to remove the equipment and replace it with a Pneumatic Cam Motor (PCM) system, using spare controllers from the 1938 tube stock. The first converted train entered service on 31 March 1955, and the stock was redesignated at CO/CP Stock, since it contained cars from both batches. All of the metadyne equipment was subsequently replaced.

Despite the shortcomings which led to its demise, the metadyne system as introduced in 1936 on the O Stock trains was the first in the world to provide regenerative braking on an electric multiple unit. Acceleration was smoother than on a train which switched starting resistances, and when braking the metadyne unit returned power to the tracks, which could be used by other trains if needed. However, conditions were not always ideal, and the substations were not really designed to cope with regeneration, which meant that often the train switched to rheostatic braking, where the power was dissipated in a resistance bank. The weight of the equipment was also a serious drawback.

Gun Control

In the period immediately before the Second World War, there was increasing interest in power-operated gun controls, although military authorities were nervous of introducing a complex system which would have to be maintained in the field. However, with the increasing speed of aircraft, the need to enable searchlights, anti-aircraft guns and naval guns to move ever faster to track their movement meant the some form of powered control was essential. Engineers were faced with the problem of making a heavy piece of equipment, such as a gun on its mounting carriage, track an input signal in a smooth and accurate fashion, with very little lag between changes in the input and the actual position of the gun mount. The gun needed to be aimed at the target at all times, and moving at the correct velocity to remain so.

A human operator anticipates errors, and can also compensate for known lags in the operation of the system. Mimicking of this behaviour had been achieved for electronic signals and low-power electromechanical systems, but gun control was on a completely different scale, with machinery weighing tons and having significant inertia needing to move at velocities of up to 30 degrees per second, and accelerations of 10 degrees per second^2. In 1937, the Admiralty had placed an order with Metropolitan Vickers for a control system for an eight-barrel Pom-Pom gun. Pestarini had developed a similar system for the Italian navy. The original design used a single Metadyne to supply a constant current to the armatures of motors mounted on several guns. Each was then controlled by manually adjusting the field current. Tustin, who did most of the design work, found that the system had a large time constant, due to the inductance of the field windings. In order to improve its response, he supplied the field windings with a constant current, and used a partially compensated Metadyne to control the armature current of each motor. Tustin compared Ward Leonard control systems, Metadynes and Amplidynes, and accepted that each had its merits, but favoured the Metadyne, of which he had several years experience from their use in traction control.

DC Motor

A DC motor is any of a class of rotary electrical machines that converts direct current electrical energy into mechanical energy. The most common types rely on the forces produced by magnetic fields. Nearly all types of DC motors have some internal mechanism, either electromechanical or electronic, to periodically change the direction of current flow in part of the motor.

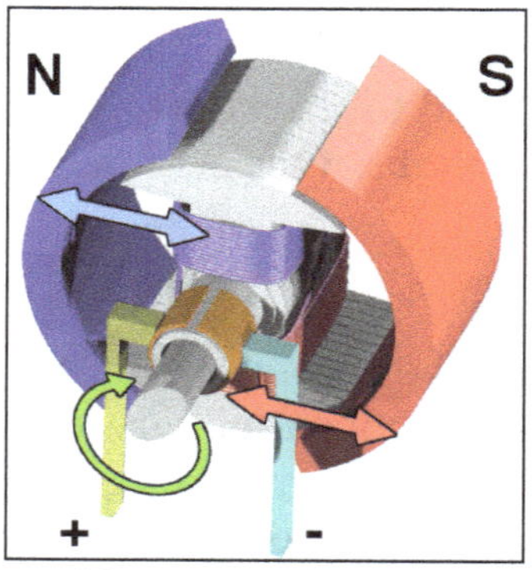

Workings of a brushed electric motor with a two-pole rotor (armature) and permanent magnet stator. "N" and "S" designate polarities on the inside axis faces of the magnets; the outside faces have opposite polarities. The + and - signs show where the DC current is applied to the commutator which supplies current to the armature coils.

The Pennsylvania Railroad's class DD1 locomotive running gear was a semi-permanently coupled pair of third rail direct current electric locomotive motors built for the railroad's initial New York-area electrification when steam locomotives were banned in the city (locomotive cab removed here).

DC motors were the first type widely used, since they could be powered from existing direct-current lighting power distribution systems. A DC motor's speed can be controlled over a wide range, using either a variable supply voltage or by changing the strength of current in its field windings. Small DC motors are used in tools, toys, and appliances. The universal motor can operate on direct current but is a lightweight motor used for portable power tools and appliances. Larger DC motors are used in propulsion of electric vehicles, elevator and hoists, or in drives for steel rolling mills. The advent of power electronics has made replacement of DC motors with AC motors possible in many applications.

Electromagnetic Motors

A coil of wire with a current running through it generates an electromagnetic field aligned with the center of the coil. The direction and magnitude of the magnetic field produced by the coil can be changed with the direction and magnitude of the current flowing through it.

A simple DC motor has a stationary set of magnets in the stator and an armature with one or more windings of insulated wire wrapped around a soft iron core that concentrates the magnetic field. The windings usually have multiple turns around the core, and in large motors there can be several parallel current paths. The ends of the wire winding are connected to a commutator. The commutator allows each armature coil to be energized in turn and connects the rotating coils with the external power supply through brushes. (Brushless DC motors have electronics that switch the DC current to each coil on and off and have no brushes.)

The total amount of current sent to the coil, the coil's size and what it's wrapped around dictate the strength of the electromagnetic field created.

The sequence of turning a particular coil on or off dictates what direction the effective electromagnetic fields are pointed. By turning on and off coils in sequence a rotating magnetic field can be created. These rotating magnetic fields interact with the magnetic fields of the magnets (permanent or electromagnets) in the stationary part of the motor (stator) to create a force on the armature which causes it to rotate. In some DC motor designs the stator fields use electromagnets to create their magnetic fields which allow greater control over the motor.

At high power levels, DC motors are almost always cooled using forced air.

Different number of stator and armature fields as well as how they are connected provide different inherent speed/torque regulation characteristics. The speed of a DC motor can be controlled by changing the voltage applied to the armature. The introduction of variable resistance in the armature circuit or field circuit allowed speed control. Modern DC motors are often controlled by power electronics systems which adjust the voltage by "chopping" the DC current into on and off cycles which have an effective lower voltage.

Since the series-wound DC motor develops its highest torque at low speed, it is often used in traction applications such as electric locomotives, and trams. The DC motor was the mainstay of electric traction drives on both electric and diesel-electric locomotives, street-cars/trams and diesel electric drilling rigs for many years. The introduction of DC motors and an electrical grid system to run machinery starting in the 1870s started a new second Industrial Revolution. DC motors can operate directly from rechargeable batteries, providing the motive power for the first electric vehicles and today's hybrid cars and electric cars as well as driving a host of cordless tools. Today DC motors are still found in applications as small as toys and disk drives, or in large sizes to operate steel rolling mills and paper machines. Large DC motors with separately excited fields were generally used with winder drives for mine hoists, for high torque as well as smooth speed control using thyristor drives. These are now replaced with large AC motors with variable frequency drives.

If external power is applied to a DC motor it acts as a DC generator, a dynamo. This feature is used to slow down and recharge batteries on hybrid car and electric cars or to return electricity back to the electric grid used on a street car or electric powered train line when they slow down. This process is called regenerative braking on hybrid and electric cars. In diesel electric locomotives they also use their DC motors as generators to slow down but dissipate the energy in resistor stacks. Newer designs are adding large battery packs to recapture some of this energy.

Brushed

The brushed DC electric motor generates torque directly from DC power supplied to the motor by using internal commutation, stationary magnets (permanent or electromagnets), and rotating electrical magnets.

Advantages of a brushed DC motor include low initial cost, high reliability, and simple control of motor speed. Disadvantages are high maintenance and low life-span for high intensity uses. Maintenance involves regularly replacing the carbon brushes and springs which carry the electric current, as well as cleaning or replacing the commutator. These components are necessary for transferring electrical power from outside the motor to the spinning wire windings of the rotor inside the motor. Brushes consist of conductors.

Brushless

Typical brushless DC motors use one or more permanent magnets in the rotor and electromagnets on the motor housing for the stator. A motor controller converts DC to AC. This design is mechanically simpler than that of brushed motors because it eliminates the complication of transferring power from outside the motor to the spinning rotor. The motor controller can sense the rotor's position via Hall effect sensors or similar devices and can precisely control the timing, phase, etc., of the current in the rotor coils to optimize torque, conserve power, regulate speed, and even apply some braking. Advantages of brushless motors include long life span, little or no maintenance, and high efficiency. Disadvantages include high initial cost, and more complicated motor speed controllers. Some such brushless motors are sometimes referred to as "synchronous motors" although they have no external power supply to be synchronized with, as would be the case with normal AC synchronous motors.

Uncommutated

Other types of DC motors require no commutation:

- Homopolar motor – A homopolar motor has a magnetic field along the axis of rotation and an electric current that at some point is not parallel to the magnetic field. The name homopolar refers to the absence of polarity change. Homopolar motors necessarily have a single-turn coil, which limits them to very low voltages. This has restricted the practical application of this type of motor.

- Ball bearing motor – A ball bearing motor is an unusual electric motor that consists of two ball bearing-type bearings, with the inner races mounted on a common conductive shaft, and the outer races connected to a high current, low voltage power supply. An alternative construction fits the outer races inside a metal tube, while the inner races are mounted on a shaft with a non-conductive section (e.g. two sleeves on an insulating rod). This method has the advantage that the tube will act as a flywheel. The direction of rotation is determined by the initial spin which is usually required to get it going.

Permanent Magnet Stators

A PM motor does not have a field winding on the stator frame, instead relying on PMs to provide the magnetic field against which the rotor field interacts to produce torque. Compensating windings in series with the armature may be used on large motors to improve commutation under load. Because this field is fixed, it cannot be adjusted for speed control. PM fields (stators) are convenient in miniature motors to eliminate the power consumption of the field winding. Most larger DC motors are of the "dynamo" type, which have stator windings. Historically, PMs could not be made to retain high flux if they were disassembled; field windings were more practical to obtain the needed amount of flux. However, large PMs are costly, as well as dangerous and difficult to assemble; this favors wound fields for large machines.

To minimize overall weight and size, miniature PM motors may use high energy magnets made with neodymium or other strategic elements; most such are neodymium-iron-boron alloy. With their higher flux density, electric machines with high-energy PMs are at least competitive with all optimally designed singly fed synchronous and induction electric machines. Miniature motors resemble the structure in the illustration, except that they have at least three rotor poles (to ensure starting, regardless of rotor position) and their outer housing is a steel tube that magnetically links the exteriors of the curved field magnets.

Wound Stators

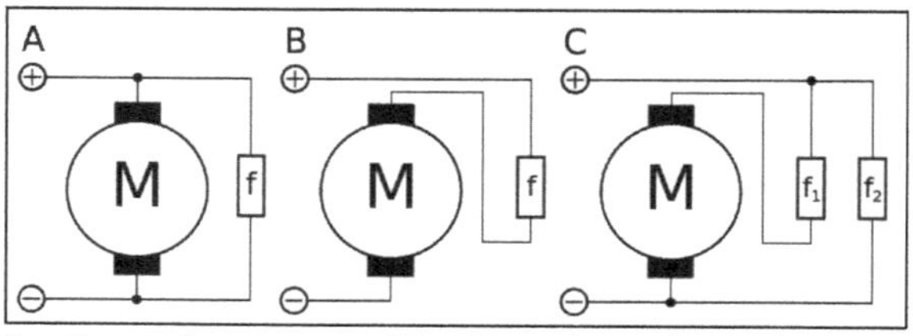

A field coil may be connected in shunt, in series, or in compound with the armature of a DC machine (motor or generator).

There are three types of electrical connections between the stator and rotor possible for DC electric motors: series, shunt/parallel and compound (various blends of series and shunt/parallel) and each has unique speed/torque characteristics appropriate for different loading torque profiles/signatures.

Series Connection

A series DC motor connects the armature and field windings in series with a common D.C. power source. The motor speed varies as a non-linear function of load torque and armature current; current is common to both the stator and rotor yielding current squared (I^2) behavior. A series motor has very high starting torque and is commonly used for starting high inertia loads, such as trains, elevators or hoists. This speed/torque characteristic is useful in applications such as dragline excavators, where the digging tool moves rapidly when unloaded but slowly when carrying a heavy load.

A series motor should never be started at no load. With no mechanical load on the series motor, the current is low, the counter-EMF produced by the field winding is weak, and so the armature must turn faster to produce sufficient counter-EMF to balance the supply voltage. The motor can be damaged by overspeed. This is called a runaway condition.

Series motors called universal motors can be used on alternating current. Since the armature voltage and the field direction reverse at the same time, torque continues to be produced in the same direction. However they run at a lower speed with lower torque on AC supply when compared to DC due to reactance voltage drop in AC which is not present in DC.Since the speed is not related to the line frequency, universal motors can develop higher-than-synchronous speeds, making them lighter than induction motors of the same rated mechanical output. This is a valuable characteristic for hand-held power tools. Universal motors for commercial utility are usually of small capacity, not more than about 1 kW output. However, much larger universal motors were used for electric locomotives, fed by special low-frequency traction power networks to avoid problems with commutation under heavy and varying loads.

Shunt Connection

A shunt DC motor connects the armature and field windings in parallel or shunt with a common D.C. power source. This type of motor has good speed regulation even as the load varies, but does not have the starting torque of a series DC motor. It is typically used for industrial, adjustable speed applications, such as machine tools, winding/unwinding machines and tensioners.

Compound Connection

A compound DC motor connects the armature and fields windings in a shunt and a series combination to give it characteristics of both a shunt and a series DC motor. This motor is used when both a high starting torque and good speed regulation is needed. The motor can be connected in two arrangements: cumulatively or differentially. Cumulative compound motors connect the series field to aid the shunt field, which provides higher starting torque but less speed regulation. Differential compound DC motors have good speed regulation and are typically operated at constant speed.

Homopolar Motor

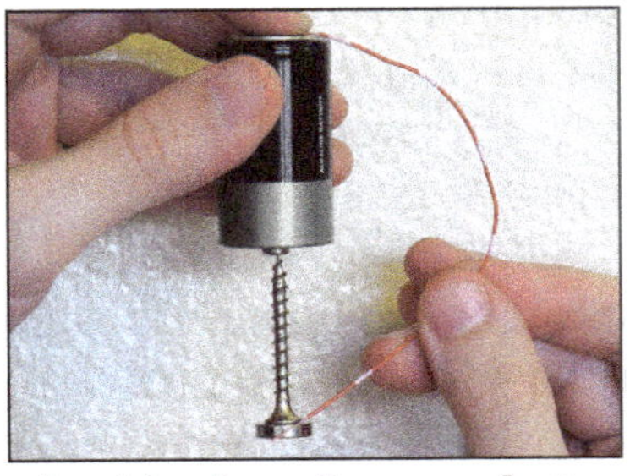

DIY simple homopolar motor made with a drywall screw, a battery cell, a wire, and a disk magnet. The magnet is on top of the screw head. The screw and magnet make contact with the bottom of the battery cell and are held together by the magnet's attraction. The screw and magnet spin, with the screw tip acting as a bearing.

A homopolar motor is a direct current electric motor with two magnetic poles, the conductors of which always cut unidirectional lines of magnetic flux by rotating a conductor around a fixed axis so that the conductor is at right angles to a static magnetic field.

The resulting EMF (Electromotive Force) being continuous in one direction, the homopolar motor needs no commutator but still requires slip rings. The name *homopolar* indicates that the electrical polarity of the conductor and the magnetic field poles do not change (i.e., that it does not require commutation).

History

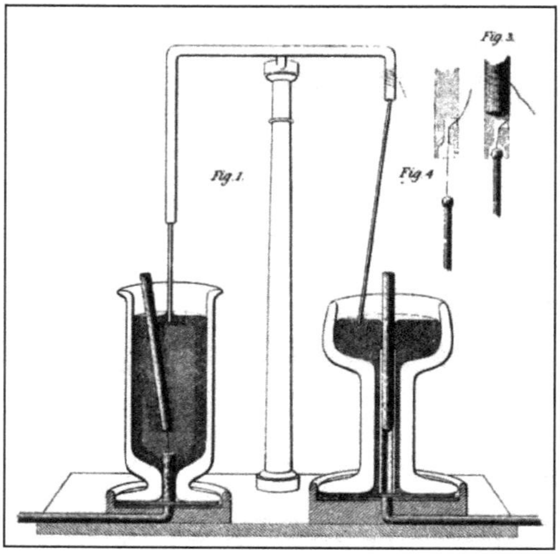

Electromagnetic rotation experiment of Faraday, *ca.* 1821.

The homopolar motor was the first electrical motor to be built. Its operation was demonstrated by Michael Faraday in 1821 at the Royal Institution in London.

In 1821, soon after the Danish physicist and chemist Hans Christian Ørsted discovered the phenomenon of electromagnetism, Humphry Davy and British scientist William Hyde Wollaston tried, but failed, to design an electric motor. Faraday, having discussed the problem with the two men, went on to build two devices to produce what he called "electromagnetic rotation". One of these, now known as the *homopolar motor*, caused a continuous circular motion that was engendered by the circular magnetic force around a wire that extended into a pool of mercury wherein was placed a magnet. The wire would then rotate around the magnet if supplied with current from a chemical battery. These experiments and inventions formed the foundation of modern electromagnetic technology. In his excitement, Faraday published results without acknowledging his work with either Wollaston or Davy. The resulting controversy within the Royal Society strained his mentor relationship with Davy and may well have contributed to Faraday's assignment to other activities, which consequently prevented his involvement in electromagnetic research for several years.

B. G. Lamme described in 1912 a homopolar machine rated 2,000 kW, 260 V, 7,700 A and 1,200 rpm with 16 slip rings operating at a peripheral velocity of 67 m/s. A unipolar generator rated 1,125 kW, 7.5 V 150,000 A, 514 rpm built in 1934 was installed in a U.S. steel mill for pipe welding purposes.

Principle of Operation

The homopolar motor is driven by the Lorentz force. A conductor with a current flowing through it when placed in a magnetic field which is perpendicular to the current feels a force in the direction perpendicular to both the magnetic field and the current. This force provides a torque around the axis of rotation. Because the axis of rotation is parallel to the magnetic field, and the opposing magnetic fields do not change polarity, no commutation is required for the conductor to keep turning. This simplicity is most readily achieved with single turn designs, which makes homopolar motors unsuitable for most practical applications.

Like most electro-mechanical machines, a homopolar motor is reversible: if the conductor is turned mechanically, then it will operate as a homopolar generator, producing a direct current voltage between the two terminals of the conductor. The direct current produced is an effect of the homopolar nature of the design. Homopolar generators (HPGs) were extensively researched in the late 20th century as low voltage but very high current DC power supplies and have achieved some success powering experimental railguns.

Building a Simple Homopolar Motor

A homopolar motor is very easy to build. A permanent magnet is used to provide the external magnetic field in which the conductor will turn, and a battery causes a current to flow along a conducting wire. It is not necessary for the magnet to move, or even to be in contact with the rest of the motor; its sole purpose is to provide a magnetic field that will interact with the magnetic field induced by the current in the wire. One can attach the magnet to the battery and allow the conducting wire to rotate freely while closing the electric circuit by touching both the top of the battery and the magnet attached to the bottom of the battery. The wire and the battery may become hot if operated continuously.

AC Motor

An industrial type of AC motor with electrical terminal box at the top and output rotating shaft on the left. Such motors are widely used for pumps, blowers, conveyors and other industrial machinery.

An AC motor is an electric motor driven by an alternating current (AC). The AC motor commonly consists of two basic parts, an outside stationary stator having coils supplied with alternating current to produce a rotating magnetic field, and an inside rotor attached to the output shaft producing a second rotating magnetic field. The rotor magnetic field may be produced by permanent magnets, reluctance saliency, or DC or AC electrical windings.

Less commonly, linear AC motors operate on similar principles as rotating motors but have their stationary and moving parts arranged in a straight line configuration, producing linear motion instead of rotation.

Operating Principles

When an AC motor is in steady-state rotation (motion), the magnetic fields of the rotor and stator rotate (move) with little or no slippage (near synchrony). The magnetic forces (repulsive and attractive) between the rotor and stator poles create average torque, capable of driving a load at rated speed. The speed of the stator rotating magnetic field (ω_s) and the speed of the rotor rotating magnetic field ( ), relative to the speed of the mechanical shaft (ω_m), must maintain synchronism for average torque production by satisfying the synchronous speed relation (i.e., $\pm\omega_s \pm \omega_r = \omega_m$). Otherwise, asynchronously rotating magnetic fields would produce pulsating or non-average torque.

The two main types of AC motors are classified as induction and synchronous. The induction motor (or asynchronous motor) always relies on a small difference in speed between the stator rotating magnetic field and the rotor shaft speed called slip to induce rotor current in the rotor AC winding. As a result, the induction motor cannot produce torque near synchronous speed where induction (or slip) is irrelevant or ceases to exist. In contrast, the synchronous motor does not rely on slip-induction for operation and uses either permanent magnets, salient poles (having projecting magnetic poles), or an independently excited rotor winding. The synchronous motor produces its rated torque at exactly synchronous speed. The brushless wound-rotor doubly fed synchronous motor system has an independently excited rotor winding that does not rely on the principles of slip-induction of current. The brushless wound-rotor doubly fed motor is a synchronous motor that can function exactly at the supply frequency or sub to super multiple of the supply frequency.

Other types of motors include eddy current motors, and also AC/DC mechanically commutated machines in which speed is dependent on voltage and winding connection.

History

Alternating current technology was rooted in Michael Faraday's and Joseph Henry's 1830-31 discovery that a changing magnetic field can induce an electric current in a circuit. Faraday is usually given credit for this discovery since he published his findings first.

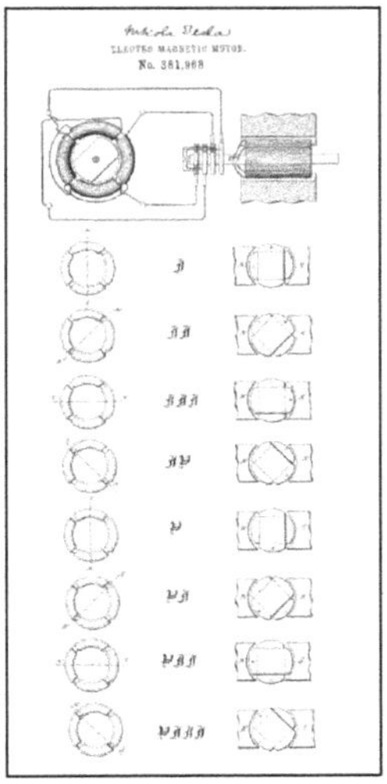

Drawing from U.S. Patent 381968, illustrating principle of Tesla's alternating current motor.

In 1832, French instrument maker Hippolyte Pixii generated a crude form of alternating current when he designed and built the first alternator. It consisted of a revolving horseshoe magnet passing over two wound wire coils.

Because of AC's advantages in long distance high voltage transmission, there were many inventors in the United States and Europe during the late 19th century trying to develop workable AC motors. The first person to conceive of a rotating magnetic field was Walter Baily, who gave a workable demonstration of his battery-operated polyphase motor aided by a commutator on June 28, 1879 to the Physical Society of London. Describing an apparatus nearly identical to Baily's, French electrical engineer Marcel Deprez published a paper in 1880 that identified the rotating magnetic field principle and that of a two-phase AC system of currents to produce it. Never practically demonstrated, the design was flawed, as one of the two currents was "furnished by the machine itself." In 1886, English engineer Elihu Thomson built an AC motor by expanding upon the induction-repulsion principle and his wattmeter. In 1887, American inventor Charles Schenk Bradley was the first to patent a two-phase AC power transmission with four wires.

"Commutatorless" alternating current induction motors seem to have been independently invented by Galileo Ferraris and Nikola Tesla. Ferraris demonstrated a working model of his single-phase induction motor in 1885, and Tesla built his working two-phase induction motor in 1887 and demonstrated it at the American Institute of Electrical Engineers in 1888 (although Tesla claimed that he conceived the rotating magnetic field in 1882). In 1888, Ferraris published his research to the Royal Academy of Sciences in Turin, where he detailed the foundations of motor operation; Tesla, in the same year, was granted a United States patent for his own motor. Working from Ferraris's experiments, Mikhail Dolivo-Dobrovolsky introduced the first three-phase induction motor in 1890, a much more capable design that became the prototype used in Europe and the U.S. He also invented the first three-phase generator and transformer and combined them into the first complete AC three-phase system in 1891. The three-phase motor design was also worked on by the Swiss engineer Charles Eugene Lancelot Brown, and other three-phase AC systems were developed by German technician Friedrich August Haselwander and Swedish engineer Jonas Wenström.

Slip

If the rotor of a squirrel cage motor were to run at the true synchronous speed, the flux in the rotor at any given place on the rotor would not change, and no current would be created in the squirrel cage. For this reason, ordinary squirrel-cage motors run at some tens of RPM slower than synchronous speed. Because the rotating field (or equivalent pulsating field) effectively rotates faster than the rotor, it could be said to *slip* past the surface of the rotor. The difference between synchronous speed and actual speed is called slip, and loading the motor increases the amount of slip as the motor slows down slightly. Even with no load, internal mechanical losses prevent the slip from being zero.

The speed of the AC motor is determined primarily by the frequency of the AC supply and the number of poles in the stator winding, according to the relation:

$$N_s = 120F / p$$

where,

N_s = Synchronous speed, in revolutions per minute.

F = AC power frequency.

p = Number of poles per phase winding.

Actual RPM for an induction motor will be less than this calculated synchronous speed by an amount known as *slip*, that increases with the torque produced. With no load, the speed will be very close to synchronous. When loaded, standard motors have between 2-3% slip, special motors may have up to 7% slip, and a class of motors known as *torque motors* are rated to operate at 100% slip (0 RPM/full stall).

The slip of the AC motor is calculated by:

$$S = (N_s - N_r) / N_s$$

where,

N_r = Rotational speed, in revolutions per minute.

S = Normalised Slip, 0 to 1.

As an example, a typical four-pole motor running on 60 Hz might have a nameplate rating of 1725 RPM at full load, while its calculated speed is 1800 RPM.

The speed in this type of motor has traditionally been altered by having additional sets of coils or poles in the motor that can be switched on and off to change the speed of magnetic field rotation. However, developments in power electronics mean that the frequency of the power supply can also now be varied to provide a smoother control of the motor speed.

This kind of rotor is the basic hardware for induction regulators, which is an exception of the use of rotating magnetic field as pure electrical (not electromechanical) application.

Polyphase Cage Rotor

Most common AC motors use the squirrel-cage rotor, which will be found in virtually all domestic and light industrial alternating current motors. The squirrel-cage refers to the rotating exercise cage for pet animals. The motor takes its name from the shape of its rotor "windings"- a ring at either end of the rotor, with bars connecting the rings running the length of the rotor. It is typically cast aluminum or copper poured between the iron laminates of the rotor, and usually only the end rings will be visible. The vast majority of the rotor currents will flow through the bars rather than the higher-resistance and usually varnished laminates. Very low voltages at very high currents are typical in the bars and end rings; high efficiency motors will often use cast copper to reduce the resistance in the rotor.

In operation, the squirrel-cage motor may be viewed as a transformer with a rotating secondary. When the rotor is not rotating in sync with the magnetic field, large rotor currents are induced; the large rotor currents magnetize the rotor and interact with the stator's magnetic fields to bring the rotor almost into synchronization with the stator's field. An unloaded squirrel-cage motor at rated no-load speed will consume electrical power only to maintain rotor speed against friction and resistance losses. As the mechanical load increases, so will the electrical load - the electrical load is inherently related to the mechanical load. This is similar to a transformer, where the primary's electrical load is related to the secondary's electrical load.

This is why a squirrel-cage blower motor may cause household lights to dim upon starting, but does not dim the lights on startup when its fan belt (and therefore mechanical load) is removed. Furthermore, a stalled squirrel-cage motor (overloaded or with a jammed shaft) will consume current limited only by circuit resistance as it attempts to start. Unless something else limits the current (or cuts it off completely) overheating and destruction of the winding insulation is the likely outcome.

Virtually every washing machine, dishwasher, standalone fan, record player, etc. uses some variant of a squirrel-cage motor.

Polyphase Wound Rotor

An alternate design, called the wound rotor, is used when variable speed is required. In this case, the rotor has the same number of poles as the stator and the windings are made of wire, connected to slip rings on the shaft. Carbon brushes connect the slip rings to a controller such as a variable resistor that allows changing the motor's slip rate. In certain high-power variable-speed wound rotor drives, the slip-frequency energy is captured, rectified, and returned to the power supply through an inverter. With bidirectionally controlled power, the wound rotor becomes an active participant in the energy conversion process, with the wound rotor doubly fed configuration showing twice the power density.

Compared to squirrel cage rotors, wound rotor motors are expensive and require maintenance of the slip rings and brushes, but they were the standard form for variable speed control before the advent of compact power electronic devices. Transistorized inverters with variable-frequency drive can now be used for speed control, and wound rotor motors are becoming less common.

Several methods of starting a polyphase motor are used. Where a large inrush current and high starting torque can be permitted, the motor can be started across the line, by applying full line voltage to the terminals (direct-on-line, DOL). Where it is necessary to limit the starting inrush current (where the motor is large compared with the short-circuit capacity of the supply), the motor is started at reduced voltage using either series inductors, an autotransformer, thyristors, or other devices. A technique sometimes used is star-delta (YΔ) starting, where the motor coils are initially connected in star configuration for acceleration of the load, then switched to delta configuration when the load is up to speed. This technique is more common in Europe than in North America. Transistorized drives can directly vary the applied voltage as required by the starting characteristics of the motor and load.

This type of motor is becoming more common in traction applications such as locomotives, where it is known as the asynchronous traction motor.

Two-phase Servo Motor

A typical two-phase AC servo-motor has a squirrel cage rotor and a field consisting of two windings:

1. A constant-voltage (AC) main winding.
2. A control-voltage (AC) winding in quadrature (i.e., 90 degrees phase shifted) with the main winding so as to produce a rotating magnetic field. Reversing phase makes the motor reverse.

An AC servo amplifier, a linear power amplifier, feeds the control winding. The electrical resistance of the rotor is made high intentionally so that the speed/torque curve is fairly linear. Two-phase servo motors are inherently high-speed, low-torque devices, heavily geared down to drive the load.

Single-phase Induction Motor

Single-phase motors do not have a unique rotating magnetic field like multi-phase motors. The field alternates (reverses polarity) between pole pairs and can be viewed as two fields rotating in opposite directions. They require a secondary magnetic field that causes the rotor to move in a specific direction. After starting, the alternating stator field is in relative rotation with the rotor. Several methods are commonly used:

Shaded-pole Motor

A common single-phase motor is the shaded-pole motor and is used in devices requiring low starting torque, such as electric fans, small pumps, or small household appliances. In this motor, small single-turn copper "shading coils" create the moving magnetic field. Part of each pole is encircled by a copper coil or strap; the induced current in the strap opposes the change of flux through the coil. This causes a time lag in the flux passing through the shading coil, so that the maximum field intensity moves higher across the pole face on each cycle. This produces a low level rotating magnetic field which is large enough to turn both the rotor and its attached load. As the rotor picks up speed the torque builds up to its full level as the principal magnetic field is rotating relative to the rotating rotor.

A reversible shaded-pole motor was made by Barber-Colman several decades ago. It had a single field coil, and two principal poles, each split halfway to create two pairs of poles. Each of these four "half-poles" carried a coil, and the coils of diagonally opposite half-poles were connected to a pair of terminals. One terminal of each pair was common, so only three terminals were needed in all.

The motor would not start with the terminals open; connecting the common to one other made the motor run one way, and connecting common to the other made it run the other way. These motors were used in industrial and scientific devices.

An unusual, adjustable-speed, low-torque shaded-pole motor could be found in traffic-light and advertising-lighting controllers. The pole faces were parallel and relatively close to each other, with the disc centred between them, something like the disc in a watthour meter. Each pole face was split, and had a shading coil on one part; the shading coils were on the parts that faced each other. Both shading coils were probably closer to the main coil; they could have both been farther away, without affecting the operating principle, just the direction of rotation.

Applying AC to the coil created a field that progressed in the gap between the poles. The plane of the stator core was approximately tangential to an imaginary circle on the disc, so the travelling magnetic field dragged the disc and made it rotate.

The stator was mounted on a pivot so it could be positioned for the desired speed and then clamped in position. Keeping in mind that the effective speed of the travelling magnetic field in the gap was constant, placing the poles nearer to the centre of the disc made it run relatively faster, and toward the edge, slower.

It is possible that these motors are still in use in some older installations.

Split-phase Motor

Another common single-phase AC motor is the *split-phase induction motor*, commonly used in major appliances such as air conditioners and clothes dryers. Compared to the shaded pole motor, these motors provide much greater starting torque.

A split-phase motor has a secondary startup winding that is 90 electrical degrees to the main winding, always centered directly between the poles of the main winding, and connected to the main winding by a set of electrical contacts. The coils of this winding are wound with fewer turns of smaller wire than the main winding, so it has a lower inductance and higher resistance. The position of the winding creates a small phase shift between the flux of the main winding and the flux of the starting winding, causing the rotor to rotate. When the speed of the motor is sufficient to overcome the inertia of the load, the contacts are opened automatically by a centrifugal switch or electric relay. The direction of rotation is determined by the connection between the main winding and the start circuit. In applications where the motor requires a fixed rotation, one end of the start circuit is permanently connected to the main winding, with the contacts making the connection at the other end.

Capacitor Start Motor

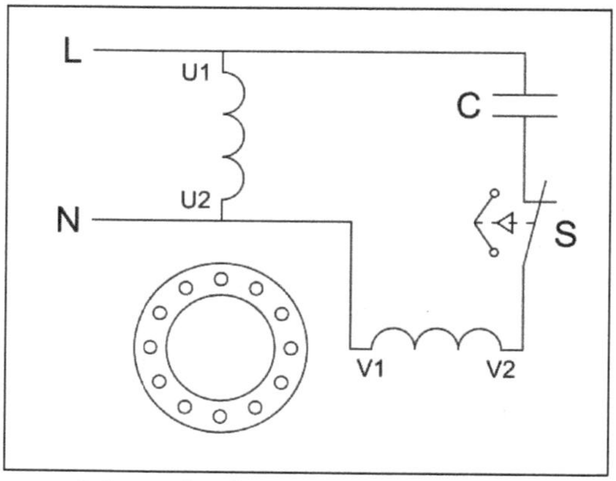

Schematic of a capacitor start motor.

A capacitor start motor is a split-phase induction motor with a starting capacitor inserted in series with the startup winding, creating an LC circuit which produces a greater phase shift (and so, a much greater starting torque) than both split-phase and shaded pole motors.

Resistance Start Motor

A resistance start motor is a split-phase induction motor with a starter inserted in series with the startup winding, creating reactance. This added starter provides assistance in the starting and initial direction of rotation.

Permanent-split Capacitor Motor

Another variation is the *permanent-split capacitor (or PSC) motor*. Also known as a capacitor-run motor, this type of motor uses a non-polarized capacitor with a high voltage rating to generate an electrical phase shift between the run and start windings. PSC motors are the dominant type of split-phase motor in Europe and much of the world, but in North America, they are most frequently used in variable torque applications (like blowers, fans, and pumps) and other cases where variable speeds are desired.

A capacitor with a relatively low capacitance, and relatively high voltage rating, is connected in series with the start winding and remains in the circuit during the entire run cycle. Like other split-phase motors, the main winding is used with a smaller start winding, and rotation is changed by reversing the connection between the main winding and the start circuit. There are significant differences, however; the use of a speed sensitive centrifugal switch requires that other split-phase motors must operate at, or very close to, full speed. PSC motors may operate within a wide range of speeds, much lower than the motor's electrical speed. Also, for applications like automatic door openers that require the motor to reverse rotation often, the use of a mechanism requires that a motor must slow to a near stop before contact with the start winding is re-established. The 'permanent' connection to the capacitor in a PSC motor means that changing rotation is instantaneous.

Three-phase motors can be converted to PSC motors by making common two windings and connecting the third via a capacitor to act as a start winding. However, the power rating needs to be at least 50% larger than for a comparable single-phase motor due to an unused winding.

Synchronous Motor

Three-phase system with rotating magnetic fields.

Polyphase Synchronous Motor

If connections to the rotor coils of a three-phase motor are taken out on slip-rings and fed a separate field current to create a continuous magnetic field (or if the rotor consists of a permanent magnet), the result is called a synchronous motor because the rotor will rotate synchronously with the rotating magnetic field produced by the polyphase electrical supply. Another synchronous motor system is the brushless wound-rotor doubly fed synchronous motor system with an independently excited rotor multiphase AC winding set that may experience slip-induction beyond synchronous speeds but like all synchronous motors, does not rely on slip-induction for torque production.

The synchronous motor can also be used as an alternator.

Nowadays, synchronous motors are frequently driven by transistorized variable-frequency drives. This greatly eases the problem of starting the massive rotor of a large synchronous motor. They may also be started as induction motors using a squirrel-cage winding that shares the common rotor: once the motor reaches synchronous speed, no current is induced in the squirrel-cage winding so it has little effect on the synchronous operation of the motor, aside from stabilizing the motor speed on load changes.

Synchronous motors are occasionally used as traction motors; the TGV may be the best-known example of such use.

Huge numbers of three phase synchronous motors are now fitted to electric cars. They have a Nd or other rare earth permanent magnet.

One use for this type of motor is its use in a power factor correction scheme. They are referred to as synchronous condensers. This exploits a feature of the machine where it consumes power at a leading power factor when its rotor is over excited. It thus appears to the supply to be a capacitor, and could thus be used to correct the lagging power factor that is usually presented to the electric supply by inductive loads. The excitation is adjusted until a near unity power factor is obtained (often automatically). Machines used for this purpose are easily identified as they have no shaft extensions. Synchronous motors are valued in any case because their power factor is much better than that of induction motors, making them preferred for very high power applications.

Some of the largest AC motors are pumped-storage hydroelectricity generators that are operated as synchronous motors to pump water to a reservoir at a higher elevation for later use to generate electricity using the same machinery. Six 500-megawatt generators are installed in the Bath County Pumped Storage Station in Virginia, USA. When pumping, each unit can produce 642,800 horsepower (479.3 megawatts).

Single-phase Synchronous Motor

Small single-phase AC motors can also be designed with magnetized rotors (or several variations on that idea).

If a conventional squirrel-cage rotor has flats ground on it to create salient poles and increase reluctance, it will start conventionally, but will run synchronously, although it can provide only a modest torque at synchronous speed. This is known as a reluctance motor.

Because inertia makes it difficult to instantly accelerate the rotor from stopped to synchronous speed, these motors normally require some sort of special feature to get started. Some include a squirrel-cage structure to bring the rotor close to synchronous speed. Various other designs use a small induction motor (which may share the same field coils and rotor as the synchronous motor) or a very light rotor with a one-way mechanism (to ensure that the rotor starts in the "forward" direction). In the latter instance, applying AC power creates chaotic jumping movement back and forth; such a motor will always start, but lacking the anti-reversal mechanism, the direction it runs is unpredictable. The Hammond organ tone generator used a non-self-starting synchronous motor (until comparatively recently), and had an auxiliary conventional shaded-pole starting motor. A spring-loaded auxiliary manual starting switch connected power to this second motor for a few seconds.

Hysteresis Synchronous Motor

These motors are relatively costly, and are used where exact speed (assuming an exact-frequency AC source) as well as rotation with a very small amount of fast variations in speed (called 'flutter" in audio recordings) is essential. Applications included tape recorder capstan drives (the motor shaft could be the capstan), and, before the advent of crystal control, motion picture cameras and recorders. Their distinguishing feature is their rotor, which is a smooth cylinder of a magnetic alloy that stays magnetized, but can be demagnetized fairly easily as well as re-magnetized with poles in a new location. Hysteresis refers to how the magnetic flux in the metal lags behind the external magnetizing force; for instance, to demagnetize such a material, one could apply a magnetizing field of opposite polarity to that which originally magnetized the material. These motors have a stator like those of capacitor-run squirrel-cage induction motors. On startup, when slip decreases sufficiently, the rotor becomes magnetized by the stator's field, and the poles stay in place. The motor then runs at synchronous speed as if the rotor were a permanent magnet. When stopped and restarted, the poles are likely to form at different locations. For a given design, torque at synchronous speed is only relatively modest, and the motor can run at below synchronous speed. In simple words, it is lagging magnetic field behind magnetic flux.

Other AC Motor Types

Universal Motor and Series Wound Motor

A universal motor is a design that can operate on either AC or DC power. In universal motors the stator and rotor of a brushed DC motor are both wound and supplied from an external source, with the torque being a function of the rotor current times the stator current so reversing the current in both rotor and stator does not reverse the rotation. Universal motors can run on AC as well as DC provided the frequency is not so high that the inductive reactance of the stator winding and/or eddy current losses become problems. Nearly all universal motors are series-wound because their stators have relatively few turns, minimizing inductance. Universal motors are compact, have high starting torque and can be varied in speed over a wide range with relatively simple controls such as rheostats and PWM choppers. Compared with induction motors, universal motors do have some drawbacks inherent to their brushes and commutators: relatively high levels of electrical and acoustic noise, low reliability and more frequent required maintenance.

Universal motors are widely used in small home appliances and hand power tools. Until the 1970s they dominated electric traction (electric, including diesel-electric railway and road vehicles); many traction power networks still use special low frequencies such as 16.7 and 25 Hz to overcome the aforementioned problems with losses and reactance. Still widely used, universal traction motors have been increasingly displaced by polyphase AC induction and permanent magnet motors with variable-frequency drives made possible by modern power semiconductor devices.

Repulsion Motor

Repulsion motors are wound-rotor single-phase AC motors that are a type of induction motor. In a repulsion motor, the armature brushes are shorted together rather than connected in series with the field, as is done with universal motors. By transformer action, the stator induces currents in the rotor, which create torque by repulsion instead of attraction as in other motors. Several types of repulsion motors have been manufactured, but the *repulsion-start induction-run* (RS-IR) motor has been used most frequently. The RS-IR motor has a centrifugal switch that shorts all segments of the commutator so that the motor operates as an induction motor once it is close to full speed. Some of these motors also lift the brushes out of contact with source voltage regulation. Repulsion motors were developed before suitable motor starting capacitors were available, and few repulsion motors are sold as of 2005.

Exterior Rotor

Where speed stability is important, some AC motors (such as some Papst motors) have the stator on the inside and the rotor on the outside to optimize inertia and cooling.

Sliding Rotor Motor

A conical rotor brake motor incorporates the brake as an integral part of the conical sliding rotor. When the motor is at rest, a spring acts on the sliding rotor and forces the brake ring against the brake cap in the motor, holding the rotor stationary. When the motor is energized, its magnetic field generates both an axial and a radial component. The axial component overcomes the spring force, releasing the brake; while the radial component causes the rotor to turn. There is no additional brake control required.

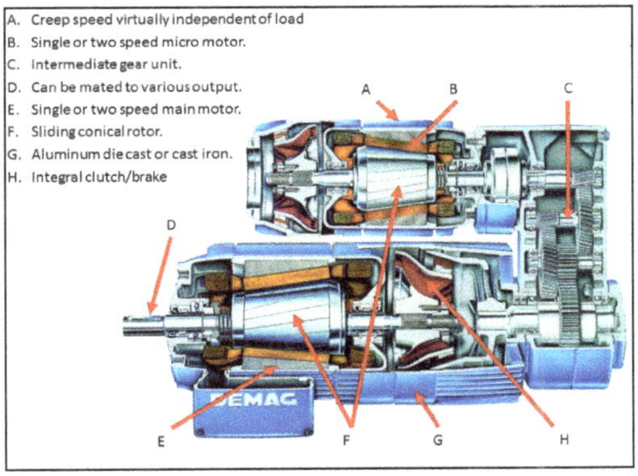

AC Motor with sliding rotors.

The high starting torque and low inertia of the conical rotor brake motor has proven to be ideal for the demands of high cycle dynamic drives in applications since the motor was invented, designed and introduced over 50 years ago. This type of motor configuration was first introduced in the USA in 1963.

Single-speed or two speed motors are designed for coupling to gear motor system gearboxes. Conical rotor brake motors are also used to power micro speed drives.

Motors of this type can also be found on overhead cranes and hoists. The micro speed unit combines two motors and an intermediate gear reducer. These are used for applications where extreme mechanical positioning accuracy and high cycling capability are needed. The micro speed unit combines a "main" conical rotor brake motor for rapid speed and a "micro" conical rotor brake motor for slow or positioning speed. The intermediate gearbox allows a range of ratios, and motors of different speeds can be combined to produce high ratios between high and low speed.

Electronically Commutated Motor

Electronically commutated (EC) motors are electric motors powered by direct-current (DC) electricity and having electronic commutation systems, rather than mechanical commutators and brushes. The current-to-torque and frequency-to-speed relationships of BLDC motors are linear. While the motor coils are powered by DC, power may be rectified from AC within the casing.

Watthour-meter Motor

These are two-phase induction motors with permanent magnets to retard the rotor so its speed is accurately proportional to the power passing through the meter. The rotor is an aluminium-alloy disc, and currents induced into it react with the field from the stator.

A split-phase watthour meter has a stator with three coils facing the disc. The magnetic circuit is completed by a C-shaped core of permeable iron. The "voltage" coil above the disc is in parallel with the supply; its many turns have a high inductance/resistance ratio (Q) so its current and magnetic field are the time integral of the applied voltage, lagging it by 90 degrees. This magnetic field passes down perpendicularly through the disc, inducing circular eddy currents in the plane of the disc centered on the field. These induced currents are proportional to the time derivative of the magnetic field, leading it by 90 degrees. This puts the eddy currents in phase with the voltage applied to the voltage coil, just as the current induced in the secondary of a transformer with a resistive load is in phase with the voltage applied to its primary.

The eddy currents pass directly above the pole pieces of two "current" coils under the disc, each wound with a few turns of heavy-gauge wire whose inductive reactance is small compared to the load impedance. These coils connect the supply to the load, producing a magnetic field in phase with the load current. This field passes from the pole of one current coil up perpendicularly through the disc and back down through the disc to the pole of the other current coil, with a completed magnetic circuit back to the first current coil. As these fields cross the disc, they pass through the eddy currents induced in it by the voltage coil producing a Lorentz force on the disc mutually perpendicular to both. Assuming power is flowing to the load, the flux from the left current coil crosses the disc upwards where the eddy current flows radially toward the center of the disc producing (by the right hand rule) a torque driving the front of the disc to the right. Similarly, the flux crosses down through the disc to the right current coil where the eddy current flows radially away from the disc center, again producing a torque driving the front of the disc to the right. When the AC polarity reverses, the eddy currents in the disc and the direction of the magnetic flux from the current coils both change, leaving the direction of the torque unchanged.

The torque is thus proportional to the instantaneous line voltage times the instantaneous load current, automatically correcting for power factor. The disc is braked by a permanent magnet so that speed is proportional to torque and the disc mechanically integrates real power. The mechanical dial on the meter reads disc rotations and the total net energy delivered to the load. (If the load supplies power to the grid, the disc rotates backwards unless prevented by a ratchet, thus making net metering possible.)

In a split-phase watthour meter the voltage coil is connected between the two "hot" (line) terminals (240V in North America) and two separate current coils are connected between the corresponding line and load terminals. No connection to the system neutral is needed to correctly handle combined line-to-neutral and line-to-line loads. Line-to-line loads draw the same current through both current coils and spin the meter twice as fast as a line-to-neutral load drawing the same current through only a single current coil, correctly registering the power drawn by the line-to-line load as twice that of the line-to-neutral load.

Other variations of the same design are used for polyphase (e.g., three-phase) power.

Slow-speed Synchronous Timing Motor

Representative are low-torque synchronous motors with a multi-pole hollow cylindrical magnet (internal poles) surrounding the stator structure. An aluminum cup supports the magnet. The stator has one coil, coaxial with the shaft. At each end of the coil are a pair of circular plates with rectangular teeth on their edges, formed so they are parallel with the shaft. They are the stator poles. One of the pair of discs distributes the coil's flux directly, while the other receives flux that has passed through a common shading coil. The poles are rather narrow, and between the poles leading from one end of the coil are an identical set leading from the other end. In all, this creates a repeating sequence of four poles, unshaded alternating with shaded, that creates a circumferential traveling field to which the rotor's magnetic poles rapidly synchronize. Some stepping motors have a similar structure.

Induction Motor

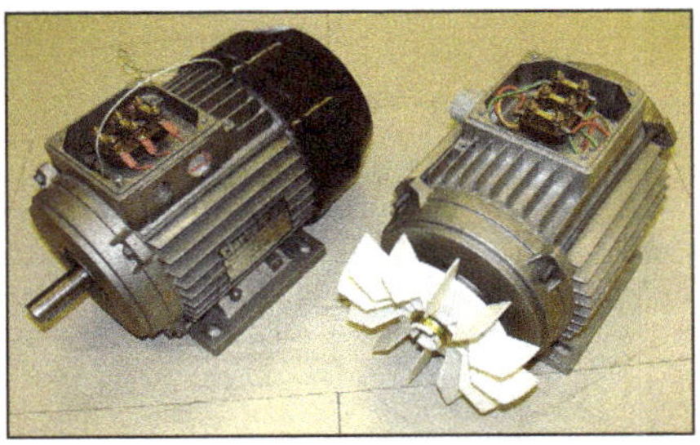

Three-phase totally enclosed fan-cooled (TEFC) induction motor with end cover on the left, and without end cover to show cooling fan. In TEFC motors, interior heat losses are dissipated indirectly through enclosure fins, mostly by forced air convection.

Cutaway view through stator of TEFC induction motor, showing rotor with internal air circulation vanes. Many such motors have a symmetric armature, and the frame may be reversed to place the electrical connection box (not shown) on the opposite side.

An induction motor, 3 phase induction motor or asynchronous motor is an AC electric motor in which the electric current in the rotor needed to produce torque is obtained by electromagnetic induction from the magnetic field of the stator winding. An induction motor can therefore be made without electrical connections to the rotor. An induction motor's rotor can be either wound type or squirrel-cage type.

Three-phase squirrel-cage induction motors are widely used in industrial drives because they are rugged, reliable and economical. Single-phase induction motors are used extensively for smaller loads, such as household appliances like fans. Although traditionally used in fixed-speed service, induction motors are increasingly being used with variable-frequency drives (VFDs) in variable-speed service. VFDs offer especially important energy savings opportunities for existing and prospective induction motors in variable-torque centrifugal fan, pump and compressor load applications. Squirrel cage induction motors are very widely used in both fixed-speed and variable-frequency drive (VFD) applications.

History

A model of Tesla's first induction motor, in Tesla Museum, Belgrade.

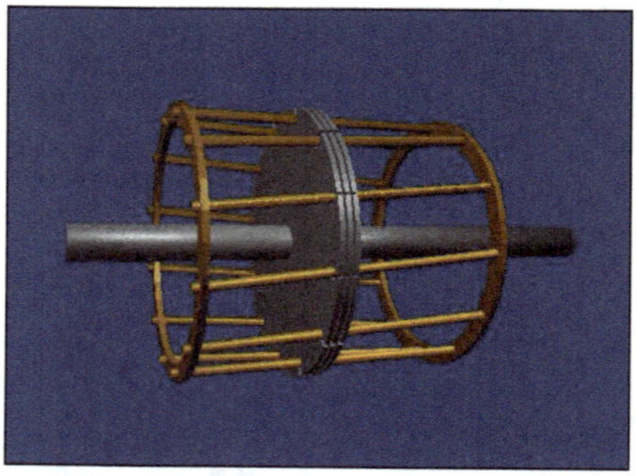

Squirrel cage rotor construction, showing only the center three laminations.

In 1824, the French physicist François Arago formulated the existence of rotating magnetic fields, termed Arago's rotations. By manually turning switches on and off, Walter Baily demonstrated this in 1879, effectively the first primitive induction motor.

The first AC commutator-free induction motors were independently invented by Galileo Ferraris and Nikola Tesla, a working motor model having been demonstrated by the former in 1885 and by the latter in 1887. Tesla applied for US patents in October and November 1887 and was granted some of these patents in May 1888. In April 1888, the *Royal Academy of Science of Turin* published Ferraris's research on his AC polyphase motor detailing the foundations of motor operation. In May 1888 Tesla presented the technical paper *A New System for Alternating Current Motors and Transformers* to the *American Institute of Electrical Engineers* (AIEE) describing three four-stator-pole motor types: one with a four-pole rotor forming a non-self-starting reluctance motor, another with a wound rotor forming a self-starting induction motor, and the third a true synchronous motor with separately excited DC supply to rotor winding.

George Westinghouse, who was developing an alternating current power system at that time, licensed Tesla's patents in 1888 and purchased a US patent option on Ferraris' induction motor concept. Tesla was also employed for one year as a consultant. Westinghouse employee C. F. Scott was assigned to assist Tesla and later took over development of the induction motor at Westinghouse. Steadfast in his promotion of three-phase development, Mikhail Dolivo-Dobrovolsky invented the cage-rotor induction motor in 1889 and the three-limb transformer in 1890. Furthermore, he claimed that Tesla's motor was not practical because of two-phase pulsations, which prompted him to persist in his three-phase work. Although Westinghouse achieved its first practical induction motor in 1892 and developed a line of polyphase 60 hertz induction motors in 1893, these early Westinghouse motors were two-phase motors with wound rotors until B. G. Lamme developed a rotating bar winding rotor.

The General Electric Company (GE) began developing three-phase induction motors in 1891. By 1896, General Electric and Westinghouse signed a cross-licensing agreement for the bar-winding-rotor design, later called the squirrel-cage rotor. Arthur E. Kennelly was the first to bring out the full significance of complex numbers (using j to represent the square root of minus one) to designate the 90° rotation operator in analysis of AC problems. GE's Charles Proteus Steinmetz greatly developed application of AC complex quantities including an analysis model now commonly known as the induction motor Steinmetz equivalent circuit.

Induction motor improvements flowing from these inventions and innovations were such that a 100-horsepower induction motor currently has the same mounting dimensions as a 7.5-horsepower motor in 1897.

Principle of Operation

In both induction and synchronous motors, the AC power supplied to the motor's stator creates a magnetic field that rotates in synchronism with the AC oscillations. Whereas a synchronous motor's rotor turns at the same rate as the stator field, an induction motor's rotor rotates at a somewhat slower speed than the stator field. The induction motor stator's magnetic field is therefore changing or rotating relative to the rotor. This induces an opposing current in the induction motor's rotor, in effect the motor's secondary winding, when the latter is short-circuited or closed through an external impedance. The rotating magnetic flux induces currents in the windings of the rotor; in a manner similar to currents induced in a transformer's secondary winding(s).

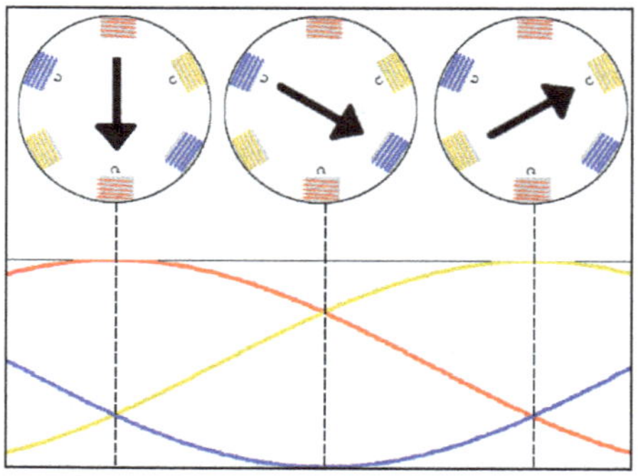

A three-phase power supply provides a rotating magnetic field in an induction motor.

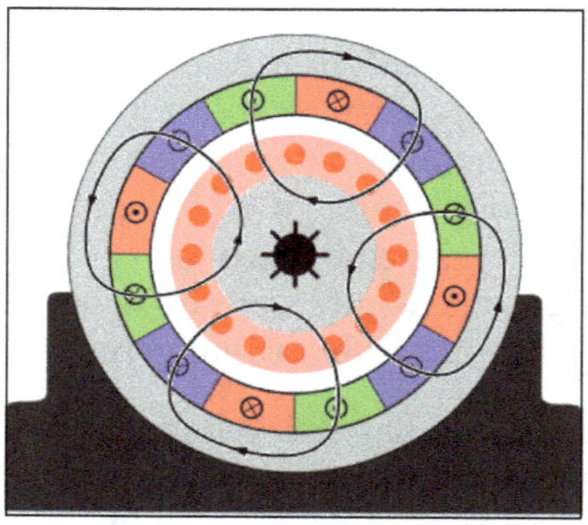

Inherent slip - unequal rotation frequency of stator field and the rotor.

The induced currents in the rotor windings in turn create magnetic fields in the rotor that react against the stator field. Due to Lenz's Law, the direction of the magnetic field created will be such as to oppose the change in current through the rotor windings. The cause of induced current in the rotor windings is the rotating stator magnetic field, so to oppose the change in rotor-winding currents the rotor will start to rotate in the direction of the rotating stator magnetic field. The rotor accelerates until the magnitude of induced rotor current and torque balances the applied mechanical load on the rotation of the rotor. Since rotation at synchronous speed would result in no induced rotor current, an induction motor always operates slightly slower than synchronous speed. The difference, or "slip," between actual and synchronous speed varies from about 0.5 to 5.0% for standard Design B torque curve induction motors. The induction motor's essential character is that it is created solely by induction instead of being separately excited as in synchronous or DC machines or being self-magnetized as in permanent magnet motors.

For rotor currents to be induced, the speed of the physical rotor must be lower than that of the stator's rotating magnetic field (n_s); otherwise the magnetic field would not be moving relative to the rotor conductors and no currents would be induced. As the speed of the rotor drops below synchronous speed, the rotation rate of the magnetic field in the rotor increases, inducing more current in the windings and creating more torque. The ratio between the rotation rate of the magnetic field induced in the rotor and the rotation rate of the stator's rotating field is called "slip". Under load, the speed drops and the slip increases enough to create sufficient torque to turn the load. For this reason, induction motors are sometimes referred to as "asynchronous motors".

An induction motor can be used as an induction generator, or it can be unrolled to form a linear induction motor which can directly generate linear motion.

Synchronous speed

An AC motor's synchronous speed, n_s, is the rotation rate of the stator's magnetic field,

$$n_s = \frac{2f}{p},$$

where f is the motor supply's frequency, where p is the number of magnetic poles and where n_s and f have identical units. For f in unit Hertz and n_s in RPM, the formula becomes,

$$n_s = \frac{2f}{p} \cdot \left(\frac{60\ s}{min} \right) = \frac{120f}{p} \cdot \left(\frac{s}{min} \right).$$

For example, for a four-pole three-phase motor, $p = 4$ and $n_s = \frac{120f}{4} = 1{,}500$ and $1{,}800$, RPM synchronous speed, respectively, for 50 Hz and 60 Hz supply systems.

The two figures at right and left above each illustrate a 2-pole 3-phase machine consisting of three pole-pairs with each pole set 60° apart.

Slip

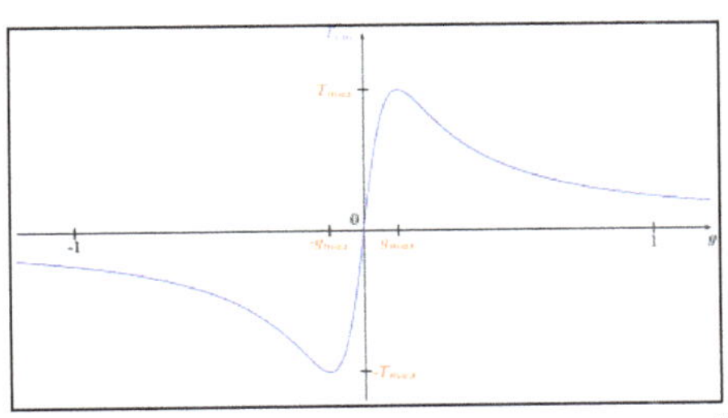

Typical torque curve as a function of slip, represented as "g" here.

Slip, s is defined as the difference between synchronous speed and operating speed, at the same frequency, expressed in rpm, or in percentage or ratio of synchronous speed.

Thus,

$$s = \frac{n_s - n_r}{n_s}$$

where n_s is stator electrical speed, n_r is rotor mechanical speed. Slip, which varies from zero at synchronous speed and 1 when the rotor is at rest, determines the motor's torque. Since the short-circuited rotor windings have small resistance, even a small slip induces a large current in the rotor and produces significant torque. At full rated load, slip varies from more than 5% for small or special purpose motors to less than 1% for large motors. These speed variations can cause load-sharing problems when differently sized motors are mechanically connected. Various methods are available to reduce slip, VFDs often offering the best solution.

Torque

Standard Torque

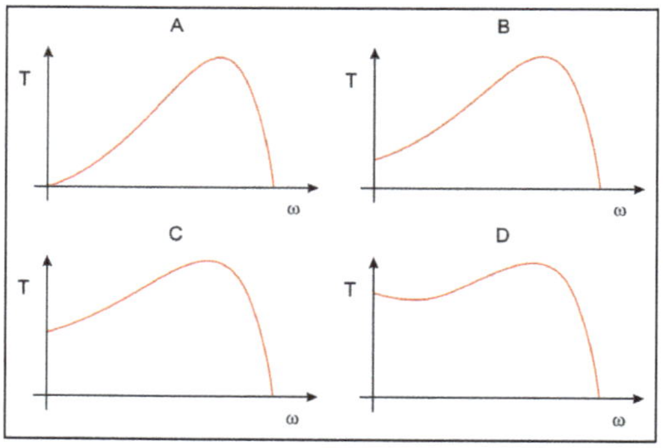

Speed-torque curves for four induction motor types: A) Single-phase, B) Polyphase cage, C) Polyphase cage deep bar, D) Polyphase double cage.

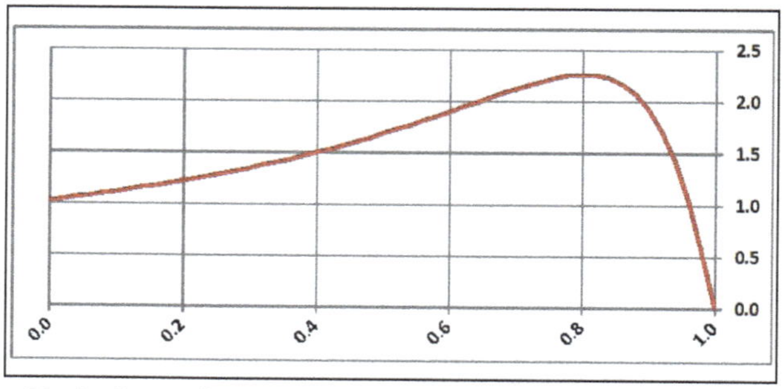

Typical speed-torque curve for NEMA Design B Motor.

The typical speed-torque relationship of a standard NEMA Design B polyphase induction motor is as shown in the curve at right. Suitable for most low performance loads such as centrifugal pumps and fans, Design B motors are constrained by the following typical torque ranges:

- *Breakdown torque* (peak torque), 175-300% of rated torque.

- *Locked-rotor torque* (torque at 100% slip), 75-275% of rated torque.
- Pull-up torque, 65-190% of rated torque.

Over a motor's normal load range, the torque's slope is approximately linear or proportional to slip because the value of rotor resistance divided by slip, R_r'/s, dominates torque in linear manner. As load increases above rated load, stator and rotor leakage reactance factors gradually become more significant in relation to R_r'/s, such that torque gradually curves towards breakdown torque. As the load torque increases beyond breakdown torque the motor stalls.

Starting

There are three basic types of competing small induction motors: single-phase, split-phase and shaded-pole types and small polyphase motors.

In two-pole single-phase motors, the torque goes to zero at 100% slip (zero speed), so these require alterations to the stator such as shaded-poles to provide starting torque. A single phase induction motor requires separate starting circuitry to provide a rotating field to the motor. The normal running windings within such a single-phase motor can cause the rotor to turn in either direction, so the starting circuit determines the operating direction.

In certain smaller single-phase motors, starting is done by means of a shaded pole with a copper wire turn around part of the pole. The current induced in this turn lags behind the supply current, creating a delayed magnetic field around the shaded part of the pole face. This imparts sufficient rotational field energy to start the motor. These motors are typically used in applications such as desk fans and record players, as the required starting torque is low, and the low efficiency is tolerable relative to the reduced cost of the motor and starting method compared to other AC motor designs.

Larger single phase motors are split-phase motors and have a second stator winding fed with out-of-phase current; such currents may be created by feeding the winding through a capacitor or having it receive different values of inductance and resistance from the main winding. In *capacitor-start* designs, the second winding is disconnected once the motor is up to speed, usually either by a centrifugal switch acting on weights on the motor shaft or a thermistor which heats up and increases its resistance, reducing the current through the second winding to an insignificant level. The *capacitor-run* designs keep the second winding on when running, improving torque. A *resistance start* design uses a starter inserted in series with the startup winding, creating reactance.

Self-starting polyphase induction motors produce torque even at standstill. Available squirrel cage induction motor starting methods include direct-on-line starting, reduced-voltage reactor or auto-transformer starting, star-delta starting or, increasingly, new solid-state soft assemblies and, of course, VFDs.

Polyphase motors have rotor bars shaped to give different speed-torque characteristics. The current distribution within the rotor bars varies depending on the frequency of the induced current. At standstill, the rotor current is the same frequency as the stator current, and tends to travel at the outermost parts of the cage rotor bars (by skin effect). The different bar shapes can give usefully different speed-torque characteristics as well as some control over the inrush current at startup.

Although polyphase motors are inherently self-starting, their starting and pull-up torque design limits must be high enough to overcome actual load conditions.

In wound rotor motors, rotor circuit connection through slip rings to external resistances allows change of speed-torque characteristics for acceleration control and speed control purposes.

Speed Control

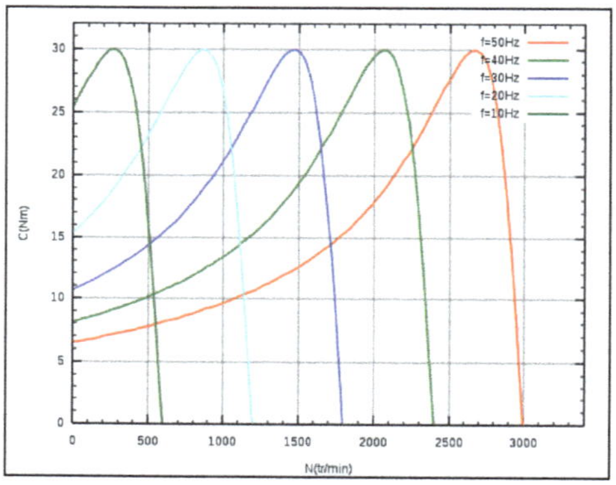

Typical speed-torque curves for different motor input frequencies as for example used with variable-frequency drives.

Before the development of semiconductor power electronics, it was difficult to vary the frequency, and cage induction motors were mainly used in fixed speed applications. Applications such as electric overhead cranes used DC drives or wound rotor motors (WRIM) with slip rings for rotor circuit connection to variable external resistance allowing considerable range of speed control. However, resistor losses associated with low speed operation of WRIMs is a major cost disadvantage, especially for constant loads. Large slip ring motor drives, termed slip energy recovery systems, some still in use, recover energy from the rotor circuit, rectify it, and return it to the power system using a VFD.

In many industrial variable-speed applications, DC and WRIM drives are being displaced by VFD-fed cage induction motors. The most common efficient way to control asynchronous motor speed of many loads is with VFDs. Barriers to adoption of VFDs due to cost and reliability considerations have been reduced considerably over the past three decades such that it is estimated that drive technology is adopted in as many as 30-40% of all newly installed motors.

Construction

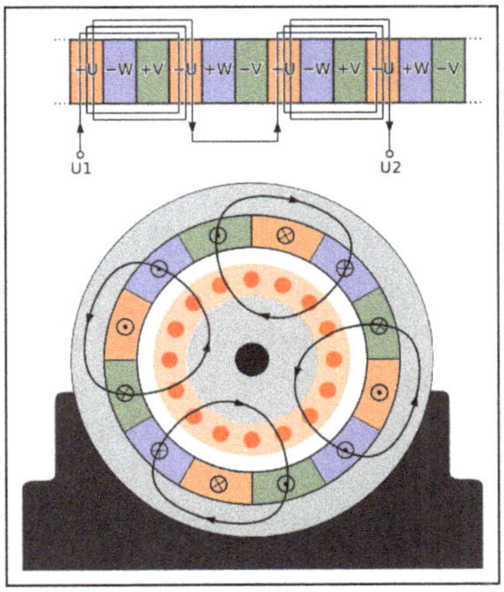

Typical winding pattern for a three-phase (U, V, W), four-pole motor.
Note the interleaving of the pole windings and the resulting quadrupole field.

The stator of an induction motor consists of poles carrying supply current to induce a magnetic field that penetrates the rotor. To optimize the distribution of the magnetic field, windings are distributed in slots around the stator, with the magnetic field having the same number of north and south poles. Induction motors are most commonly run on single-phase or three-phase power, but two-phase motors exist; in theory, induction motors can have any number of phases. Many single-phase motors having two windings can be viewed as two-phase motors, since a capacitor is used to generate a second power phase 90° from the single-phase supply and feeds it to the second motor winding. Single-phase motors require some mechanism to produce a rotating field on startup. Cage induction motor rotor's conductor bars are typically skewed to avoid magnetic locking.

Standardized NEMA & IEC motor frame sizes throughout the industry result in interchangeable dimensions for shaft, foot mounting, general aspects as well as certain motor flange aspect. Since an open, drip proof (ODP) motor design allows a free air exchange from outside to the inner stator windings, this style of motor tends to be slightly more efficient because the windings are cooler. A lower speed requires a larger frame.

Rotation Reversal

The method of changing the direction of rotation of an induction motor depends on whether it is a three-phase or single-phase machine. In the case of three-phase, reversal is straightforwardly implemented by swapping connection of any two phase conductors.

In a single-phase split-phase motor, reversal is achieved by changing the connection between the primary winding and the start circuit. Some single-phase split-phase motors that are designed for specific applications may have the connection between the primary winding and the start circuit connected internally so that the rotation cannot be changed. Also, single-phase shaded-pole motors have a fixed rotation, and the direction cannot be changed except by disassembly of the motor and reversing the stator to face opposite relative to the original rotor direction.

Power Factor

The power factor of induction motors varies with load, typically from around 0.85 or 0.90 at full load to as low as about 0.20 at no-load, due to stator and rotor leakage and magnetizing reactances. Power factor can be improved by connecting capacitors either on an individual motor basis or, by preference, on a common bus covering several motors. For economic and other considerations, power systems are rarely power factor corrected to unity power factor. Power capacitor application with harmonic currents requires power system analysis to avoid harmonic resonance between capacitors and transformer and circuit reactances. Common bus power factor correction is recommended to minimize resonant risk and to simplify power system analysis.

Efficiency

Full load motor efficiency varies from about 85% to 97%, related motor losses being broken down roughly as follows:

- Friction and windage, 5% – 15%.
- Iron or core losses, 15% – 25%.
- Stator losses, 25% – 40%.
- Rotor losses, 15% – 25%.
- Stray load losses, 10% – 20%.

Various regulatory authorities in many countries have introduced and implemented legislation to encourage the manufacture and use of higher efficiency electric motors. There is existing and forthcoming legislation regarding the future mandatory use of premium-efficiency induction-type motors in defined equipment.

Steinmetz Equivalent Circuit

Many useful motor relationships between time, current, voltage, speed, power factor and torque can be obtained from analysis of the Steinmetz equivalent circuit (also termed T-equivalent circuit or IEEE recommended equivalent circuit), a mathematical model used to describe how an induction motor's electrical input is transformed into useful mechanical energy output. The equivalent circuit is a single-phase representation of a multiphase induction motor that is valid in steady-state balanced-load conditions.

The Steinmetz equivalent circuit is expressed simply in terms of the following components:

- Stator resistance and leakage reactance (R_s, X_s).
- Rotor resistance, leakage reactance, and slip (R_r, X_r or R'_r, X'_r, and s).
- Magnetizing reactance (X_m).

Paraphrasing from Alger in Knowlton, an induction motor is simply an electrical transformer the magnetic circuit of which is separated by an air gap between the stator winding and the moving rotor winding. The equivalent circuit can accordingly be shown either with equivalent circuit components of respective windings separated by an ideal transformer or with rotor components referred to the stator side as shown in the following circuit and associated equation and parameter definition tables.

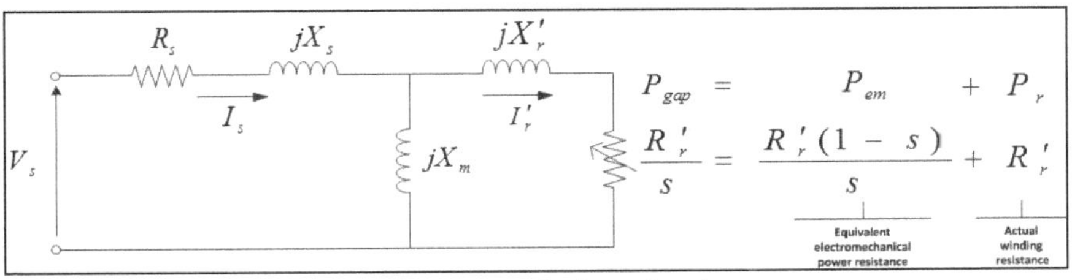

Steinmetz equivalent circuit.

Circuit Parameter Definitions		
		Units
f_s	stator synchronous frequency	Hz
n_r	rotor speed in revolutions per minute	rpm
n_s	synchronous speed in revolutions per minute	rpm
I_s	stator or primary current	A
I'_r	rotor or secondary current referred to stator side	A
I_m	magnetizing current	A
$j=\sqrt{-1}$	imaginary number, or 90° rotation, operator	
K_{TE}	$=X_m/(X_s+X_m)$ Thévenin reactance factor	
m	number of motor phases	
p	number of motor poles	
P_{em}	electromechanical power	W or hp

P_{gap}	air gap power	W
P_r	rotor copper losses	W
P_o	input power	W
P_h	core loss	W
P_f	friction and windage loss	W
P_{rl}	running light watts input	W
P_{sl}	stray-load loss	W
R_s, X_s	stator or primary resistance and leakage reactance	Ω
R'_r, X'_r	rotor or secondary resistance & leakage reactance referred to the stator side	Ω
R_o, X_o	resistance & leakage reactance at motor input	Ω
R_{TE}, X_{TE}	Thévenin equivalent resistance & leakage reactance combining R_s, X_s and X_m	Ω
s	slip	
T_{em}	electromagnetic torque	Nm or ft.-lb.
T_{max}	breakdown torque	Nm or ft.-lb.
V_s	impressed stator phase voltage	V
X_m	magnetizing reactance	Ω
X	$X_s + X'_r$	Ω
Z_s	stator or primary impedance	Ω
Z'_r	rotor or secondary impedance referred to the primary	Ω
Z_o	impedance at motor stator or primary input	Ω
Z	combined rotor or secondary and magnetizing impedance	Ω
TE	Thévenin equivalent circuit impedance, $R_{TE} + X_{TE}$	Ω
ω_r	rotor speed	rad/s

ω_s	synchronous speed	rad/s
Y	$= G - jB = \frac{1}{Z} = \frac{1}{R + jX} = \frac{R}{Z^2} - \frac{jX}{Z^2}$	mho
$\lvert Z \rvert$	$\sqrt{R^2 + X^2}$	Ω

The following rule-of-thumb approximations apply to the circuit:

- Maximum current happens under locked rotor current (LRC) conditions and is somewhat less than V_s / X, with LRC typically ranging between 6 and 7 times rated current for standard Design B motors.
- Breakdown torque T_{max} happens when $s \approx R'_r / X$ and $I_s \approx 0.7LRC$ such that $T_{max} \approx K * V_s^2 / (2X)$ and thus, with constant voltage input, a low-slip induction motor's percent-rated maximum torque is about half its percent-rated LRC.
- The relative stator to rotor leakage reactance of standard Design B cage induction motors is,

$$\frac{X_s}{X'_r} \approx \frac{0.4}{0.6}.$$

- Neglecting stator resistance, an induction motor's torque curve reduces to the Kloss equation,

$$T_{em} \approx \frac{2T_{max}}{\frac{s}{s_{max}} + \frac{s_{max}}{s}},$$ where s_{max} is slip at T_{max}.

Basic Electrical Equations,

$$\omega_s = \frac{2\pi n_s}{60} = \frac{4\pi f_s}{p}$$

Motor input equivalent impedance,

$$Z_m = R_s + jX_s + \frac{(\frac{R'_r}{s} + jX'_r)(jX_m)}{\frac{R'_r}{s} + j(X'_r + X_m)}$$

Stator current,

$$I_s = V_s / Z_m = V_s / R_s + jX_s + \frac{(\frac{R'_r}{s} + jX'_r)(jX_m)}{\frac{R'_r}{s} + j(X'_r + X_m)}$$

Rotor current referred to the stator side in terms of stator current,

$$I'_r = \frac{jX_m}{\frac{R'_r}{s} + j(X'_r + X_m)} I_s$$

Power Equations,

From Steinmetz equivalent circuit, we have,

$$\frac{R'_r}{s} = \frac{R'_r(1-s)}{s} + R'_r$$

That is, air gap power is equal to electromechanical power output plus rotor copper losses,

$$P_{gap} = P_{em} + P_r$$

$$P_r = 3R'_r I'^2_r$$

$$P_{gap} = \frac{3R'_r I'^2_r}{s}$$

$$P_{em} = \frac{3R'_r I'^2_r (1-s)}{s}$$

$$P_{em} = P_{gap}(1-s)$$

Expressing electromechanical power output in terms of rotor speed,

$$P_{em} = \frac{3R'_r I'^2_r n_r}{sn_s} \text{ (watts)}$$

$$P_{em} = \frac{3R'_r I'^2_r n_r}{746sn_s} \text{ (hp)}$$

Expressing T_{em} in ft.-lb.:

$$P_{em} = \frac{T_{em} n_r}{5252} \text{ (hp)}$$

Torque Equations,

$$T_{em} = \frac{P_{em}}{\omega_r} = \frac{\frac{P_r}{s}}{\omega_s} = \frac{3I_r'^2 R_r'}{\omega_s s} \text{ (newton-meters)}$$

In order to be able to express T_{em} directly in terms of s, IEEE recommends that R_s, X_s and be converted to the Thévenin equivalent circuit,

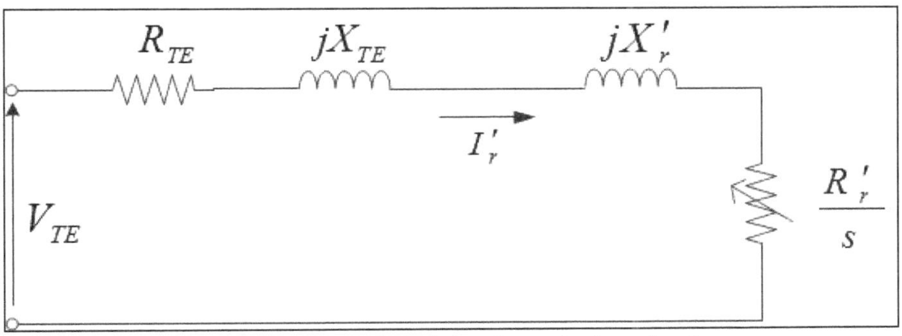

IEEE recommended Thévenin equivalent circuit.

where,

$$V_{TE} = \frac{X_m}{\sqrt{R_s^2 + (X_s + X_m)^2}} V_s$$

$$Z_{TE} = R_{TE} + jX_{TE} = \frac{jX_m (R_s + jX_s)}{R_s + j(X_s + X_m)}$$

Since $R_s^2 \gg (X_s + X_m)^2$ and $X_s \ll X_m$, and letting $K_{TE} = \frac{X_m}{X_s + X_m}$

$V_{TE} \approx Z_{TE} V_s$ and $Z_{TE} \approx K_{TE}^2 R_s + jX_s$

$$T_{em} = \frac{3V_{TE}^2}{(R_{TE} + \frac{R_r'}{s})^2 + (X_{TE} + X_r')^2} . \frac{R_r'}{s} . \frac{1}{\omega_s} \text{(N.m)}$$

For low values of slip:

Since $R_{TE} + R_r' \gg R_{TE} + X_r'$ and $R_r' \gg R_{TE}$

$$T_{em} \approx \frac{1}{\omega_s} \cdot \frac{3V_{TE}^2}{R'_r} .s \text{ (N.m)}$$

For high values of slip,

Since $R_{TE} + R'_r \ll R_{TE} + X'_r$

$$T_{em} \approx \frac{1}{\omega_s} \cdot \frac{3V_{TE}^2}{(X_s + X'_r)^2} \cdot \frac{R'^2_r}{s} \text{ (N.m)}$$

For maximum or breakdown torque, which is independent of rotor resistance,

$$T_{em} \approx \frac{1}{2\omega_s} \cdot \frac{3V_{TE}^2}{R_{TE} + \sqrt{R_{TE}^2 + (X_{TE} + X'_r)^2}} \text{ (N.m)}$$

Corresponding slip at maximum or breakdown torque is,

$$s = \frac{R'_r}{\sqrt{R_{TE}^2 + (X_{TE} + X'_r)^2}}$$

In foot-pound units,

$$T_{em} = \frac{21.21 I'^2_r R'_r}{n_r s} \text{ (ft-lb)}$$

$$T_{em} = \frac{7.04 P_{gap}}{n_s} \text{ (ft-lb)}$$

Linear Induction Motor

Linear induction motors, which work on same general principles as rotary induction motors (frequently three-phase), are designed to produce straight line motion. Uses include magnetic levitation, linear propulsion, linear actuators, and liquid metal pumping.

Transformer

A transformer is an electrical device that transfers electrical energy between two or more circuits through electromagnetic induction. A varying current in one coil of the transformer produces a varying magnetic field, which in turn induces a voltage in a second coil. Power can be transferred between the two coils through the magnetic field, without a metallic connection between the two circuits. Faraday's law of induction discovered in 1831 described this effect. Transformers are used to increase or decrease the alternating voltages in electric power applications.

Pole-mounted distribution transformer with center-tapped secondary winding used to provide "split-phase" power for residential and light commercial service, which in North America is typically rated 120/240 V.

Since the invention of the first constant-potential transformer in 1885, transformers have become essential for the transmission, distribution, and utilization of alternating current electrical energy. A wide range of transformer designs is encountered in electronic and electric power applications. Transformers range in size from RF transformers less than a cubic centimeter in volume to units interconnecting the power grid weighing hundreds of tons.

Basic Principles

Ideal Transformer

Ideal transformer equations (eq.)

By Faraday's law of induction:

$$V_S = -N_S \frac{d\Phi}{dt} \ldots (1)$$

$$V_P = -N_P \frac{d\Phi}{dt} \ldots (2)$$

Combining ratio of (1) & (2)

$$\text{Turns ratio} = \frac{V_P}{V_S} = \frac{-N_P}{-N_S} = a \ldots (3) \text{ where,}$$

for step-down transformers, $a > 1$.

for step-up transformers, $a < 1$.

By law of conservation of energy, apparent, real and reactive power are each conserved in the input and output,

$$S = I_P V_P = I_S V_S \ldots (4)$$

Combining (3) & (4) with this end note yields the ideal transformer identity,

$$\frac{V_P}{V_S} = \frac{I_S}{I_P} = \frac{N_P}{N_S} = \sqrt{\frac{L_P}{L_S}} = a \,.\, (5)$$

By Ohm's law and ideal transformer identity,

$$Z_L = \frac{V_S}{I_S} \ldots (6)$$

Apparent load impedance Z'_L (Z_L referred to the primary),

$$Z'_L = \frac{V_P}{I_P} = \frac{aV_S}{I_S / a} = a^2 \frac{V_S}{I_S} = a^2 Z_L \,.\, (7)$$

For simplification or approximation purposes, it is very common to analyze the transformer as an ideal transformer model as presented in the two images. An ideal transformer is a theoretical, linear transformer that is lossless and perfectly coupled; that is, there are no energy losses and flux is completely confined within the magnetic core. Perfect coupling implies infinitely high core magnetic permeability and winding inductances and zero net magnetomotive force.

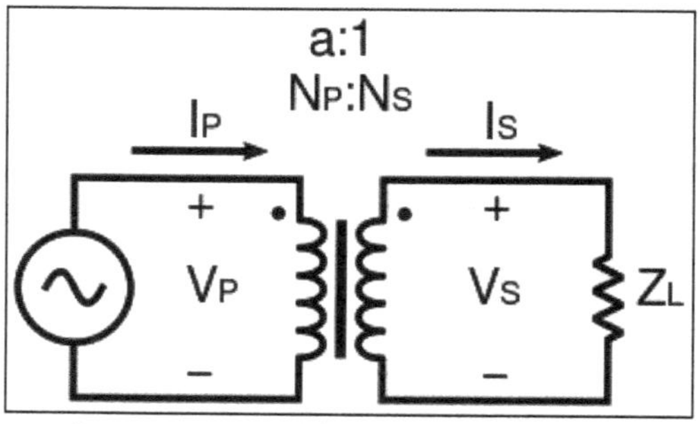

Ideal transformer connected with source V_P on primary and load impedance Z_L on secondary, where $0 < Z_L < \infty$.

A varying current in the transformer's primary winding creates a varying magnetic flux in the transformer core and a varying magnetic field impinging on the secondary winding. This varying magnetic field at the secondary winding induces a varying EMF or voltage in the secondary winding due to electromagnetic induction. The primary and secondary windings are wrapped around a core of infinitely high magnetic permeability so that all of the magnetic flux passes through both the primary and secondary windings. With a voltage source connected to the primary winding and load impedance connected to the secondary winding, the transformer currents flow in the indicated directions.

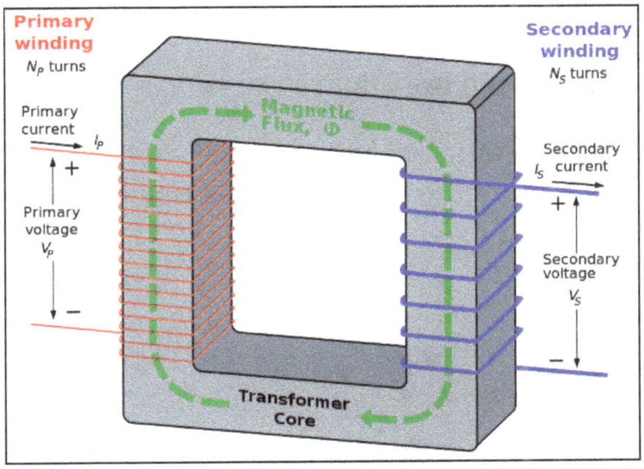

Ideal transformer and induction law.

According to Faraday's law, since the same magnetic flux passes through both the primary and secondary windings in an ideal transformer, a voltage is induced in each winding, according to eq. (1) in the secondary winding case, according to eq. (2) in the primary winding case. The primary EMF is sometimes termed counter EMF. This is in accordance with Lenz's law, which states that induction of EMF always opposes development of any such change in magnetic field.

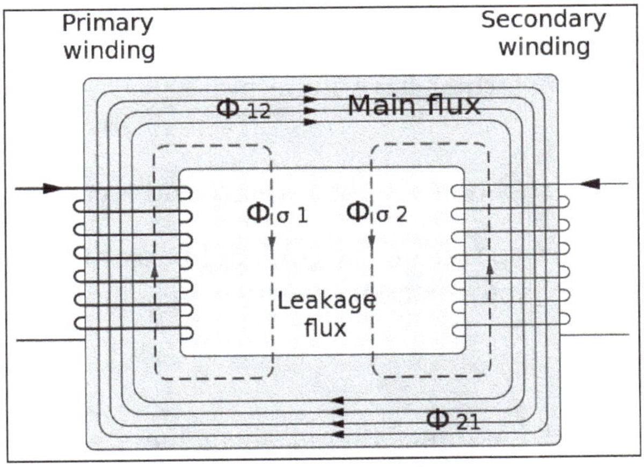

Leakage flux of a transformer.

The transformer winding voltage ratio is thus shown to be directly proportional to the winding turns ratio according to eq. (3). common usage having evolved over time from 'turn ratio' to 'turns ratio'. However, some sources use the inverse definition.

According to the law of conservation of energy, any load impedance connected to the ideal transformer's secondary winding results in conservation of apparent, real and reactive power consistent with eq. (4).

The ideal transformer identity shown in eq. (5) is a reasonable approximation for the typical commercial transformer, with voltage ratio and winding turns ratio both being inversely proportional to the corresponding current ratio.

By Ohm's law and the ideal transformer identity:

- The secondary circuit load impedance can be expressed as eq. (6),
- The apparent load impedance *referred* to the primary circuit is derived in eq. (7) to be equal to the turns ratio squared times the secondary circuit load impedance.

Real transformer

Deviations from Ideal

The ideal transformer model neglects the following basic linear aspects in real transformers:

a) Core losses, collectively called magnetizing current losses, consisting of:

- Hysteresis losses due to nonlinear application of the voltage applied in the transformer core, and
- Eddy current losses due to joule heating in the core that are proportional to the square of the transformer's applied voltage.

b) Whereas windings in the ideal model have no resistances and infinite inductances, the windings in a real transformer have finite non-zero resistances and inductances associated with:

- Joule losses due to resistance in the primary and secondary windings,
- Leakage flux that escapes from the core and passes through one winding only resulting in primary and secondary reactive impedance.

Leakage Flux

The ideal transformer model assumes that all flux generated by the primary winding links all the turns of every winding, including itself. In practice, some flux traverses paths that take it outside the windings. Such flux is termed *leakage flux*, and results in leakage inductance in series with the mutually coupled transformer windings. Leakage flux results in energy being alternately stored in and discharged from the magnetic fields with each cycle of the power supply. It is not directly a power loss, but results in inferior voltage regulation, causing the secondary voltage not to be directly proportional to the primary voltage, particularly under heavy load. Transformers are therefore normally designed to have very low leakage inductance.

In some applications increased leakage is desired, and long magnetic paths, air gaps, or magnetic bypass shunts may deliberately be introduced in a transformer design to limit the short-circuit current it will supply. Leaky transformers may be used to supply loads that exhibit negative resistance, such as electric arcs, mercury- and sodium- vapor lamps and neon signs or for safely handling loads that become periodically short-circuited such as electric arc welders.

Air gaps are also used to keep a transformer from saturating, especially audio-frequency transformers in circuits that have a DC component flowing in the windings.

Knowledge of leakage inductance is also useful when transformers are operated in parallel. It can be shown that if the percent impedance[i] and associated winding leakage reactance-to-resistance (X/R) ratio of two transformers were hypothetically exactly the same, the transformers would share power in proportion to their respective volt-ampere ratings (e.g. 500 kVA unit in parallel with 1,000 kVA unit, the larger unit would carry twice the current). However, the impedance tolerances of commercial transformers are significant. Also, the Z impedance and X/R ratio of different capacity transformers tends to vary, corresponding 1,000 kVA and 500 kVA units' values being, to illustrate, respectively, $Z \approx 5.75\%$, $X/R \approx 3.75$ and $Z \approx 5\%$, $X/R \approx 4.75$.

Equivalent Circuit

Referring to the diagram, a practical transformer's physical behavior may be represented by an equivalent circuit model, which can incorporate an ideal transformer.

Winding joule losses and leakage reactances are represented by the following series loop impedances of the model:

- Primary winding: R_P, X_P.
- Secondary winding: R_S, X_S.

In normal course of circuit equivalence transformation, R_S and X_S are in practice usually referred to the primary side by multiplying these impedances by the turns ratio squared, $(N_P/N_S)^2 = a^2$.

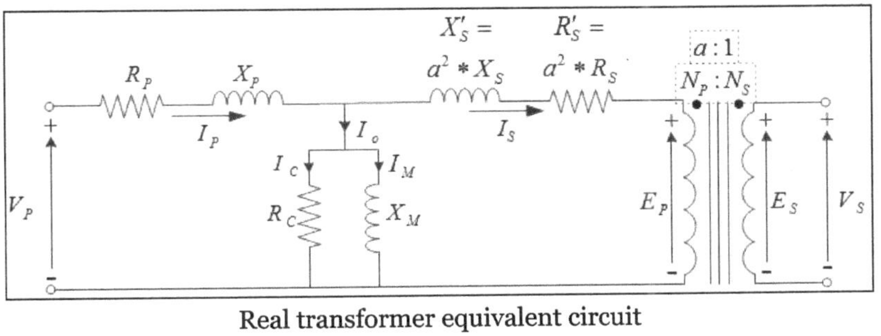

Real transformer equivalent circuit

Core loss and reactance is represented by the following shunt leg impedances of the model:

- Core or iron losses: R_C.
- Magnetizing reactance: X_M.

R_C and X_M are collectively termed the *magnetizing branch* of the model.

Core losses are caused mostly by hysteresis and eddy current effects in the core and are proportional to the square of the core flux for operation at a given frequency. The finite permeability core requires a magnetizing current I_M to maintain mutual flux in the core. Magnetizing current is in phase with the flux, the relationship between the two being non-linear due to saturation effects. However, all impedances of the equivalent circuit shown are by definition linear and such non-linearity effects are not typically reflected in transformer equivalent circuits. With sinusoidal supply, core flux lags the induced EMF by 90°. With open-circuited secondary winding, magnetizing branch current I_0 equals transformer no-load current.

Instrument transformer, with polarity dot and X1 markings on LV side terminal.

The resulting model, though sometimes termed 'exact' equivalent circuit based on linearity assumptions, retains a number of approximations. Analysis may be simplified by assuming that magnetizing branch impedance is relatively high and relocating the branch to the left of the primary impedances. This introduces error but allows combination of primary and referred secondary resistances and reactances by simple summation as two series impedances.

Transformer equivalent circuit impedance and transformer ratio parameters can be derived from the following tests: open-circuit test, short-circuit test, winding resistance test, and transformer ratio test.

Basic Transformer Parameters and Construction

Polarity

A dot convention is often used in transformer circuit diagrams, nameplates or terminal markings to define the relative polarity of transformer windings. Positively increasing instantaneous current entering the primary winding's dot end induces positive polarity voltage at the secondary winding's dot end.

Effect of Frequency

By Faraday's law of induction shown in eq. (1) and (2), transformer EMFs vary according to the derivative of flux with respect to time. The ideal transformer's core behaves linearly with time for any non-zero frequency. Flux in a real transformer's core behaves non-linearly in relation to magnetization current as the instantaneous flux increases beyond a finite linear range resulting in magnetic saturation associated with increasingly large magnetizing current, which eventually leads to transformer overheating.

Transformer Universal EMF Equation

If the flux in the core is purely sinusoidal, the relationship for either winding between its rms voltage E_{rms} of the winding, and the supply frequency f, number of turns N, core cross-sectional area a in m^2 and peak magnetic flux density B_{peak} in Wb/m^2 or T (tesla) is given by the universal EMF equation:

$$E_{rms} = \frac{2\pi f N a B_{peak}}{\sqrt{2}} \approx 4.44 f N a B_{peak}$$

If the flux does not contain even harmonics the following equation can be used for half-cycle average voltage E_{avg} of any waveshape:

$$E_{avg} = 4 f N a B_{peak}$$

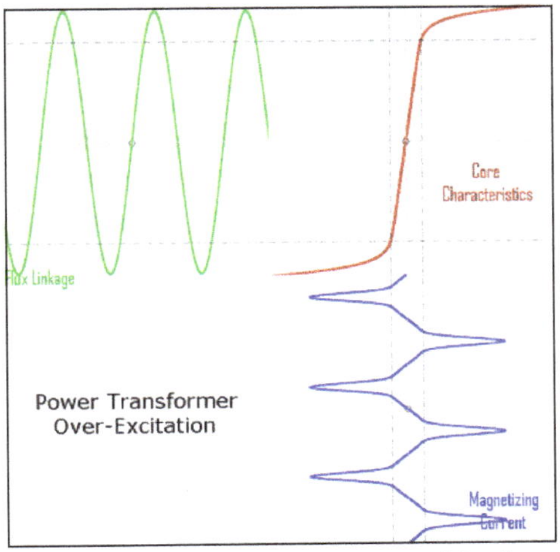

Power transformer over-excitation condition caused by decreased frequency; flux (green), iron core's magnetic characteristics (red) and magnetizing current (blue).

The EMF of a transformer at a given flux density increases with frequency. By operating at higher frequencies, transformers can be physically more compact because a given core is able to transfer more power without reaching saturation and fewer turns are needed to achieve the same impedance. However, properties such as core loss and conductor skin effect also increase with frequency. Aircraft and military equipment employ 400 Hz power supplies which reduce core and winding weight. Conversely, frequencies used for some railway electrification systems were much lower (e.g. 16.7 Hz and 25 Hz) than normal utility frequencies (50–60 Hz) for historical reasons concerned mainly with the limitations of early electric traction motors. Consequently, the transformers used to step-down the high overhead line voltages (e.g. 15 kV) were much larger and heavier for the same power rating than those required for the higher frequencies.

Operation of a transformer at its designed voltage but at a higher frequency than intended will lead to reduced magnetizing current. At a lower frequency, the magnetizing current will increase. Operation of a transformer at other than its design frequency may require assessment of voltages, losses, and cooling to establish if safe operation is practical. For example, transformers may need to be equipped with 'volts per hertz' over-excitation, ANSI function 24, relays to protect the transformer from overvoltage at higher than rated frequency.

One example is in traction transformers used for electric multiple unit and high-speed train service operating across regions with different electrical standards. The converter equipment and traction transformers have to accommodate different input frequencies and voltage (ranging from as high as 50 Hz down to 16.7 Hz and rated up to 25 kV) while being suitable for multiple AC asynchronous motor and DC converters and motors with varying harmonics mitigation filtering requirements.

Large power transformers are vulnerable to insulation failure due to transient voltages with high-frequency components, such as caused in switching or by lightning.

At much higher frequencies the transformer core size required drops dramatically: a physically small and cheap transformer can handle power levels that would require a massive iron core at mains frequency. The development of switching power semiconductor devices and complex integrated circuits made switch-mode power supplies viable, to generate a high frequency from a much lower one (or DC), change the voltage level with a small transformer, and, if necessary, rectify the changed voltage.

Energy Losses

Real transformer energy losses are dominated by winding resistance joule and core losses. Transformers' efficiency tends to improve with increasing transformer capacity. The efficiency of typical distribution transformers is between about 98 and 99 percent.

As transformer losses vary with load, it is often useful to express these losses in terms of no-load loss, full-load loss, half-load loss, and so on. Hysteresis and eddy current losses are constant at all load levels and dominate overwhelmingly without load, while variable winding joule losses dominating increasingly as load increases. The no-load loss can be significant, so that even an idle transformer constitutes a drain on the electrical supply. Designing energy efficient transformers for lower loss requires a larger core, good-quality silicon steel, or even amorphous steel for the core and thicker wire, increasing initial cost. The choice of construction represents a trade-off between initial cost and operating cost.

Transformer losses arise from:

Winding joule losses:

Current flowing through a winding's conductor causes joule heating. As frequency increases, skin effect and proximity effect causes the winding's resistance and, hence, losses to increase.

Core losses:

Hysteresis losses:

Each time the magnetic field is reversed, a small amount of energy is lost due to hysteresis within the core. According to Steinmetz's formula, the heat energy due to hysteresis is given by:

$W_h \approx \eta \beta_{max}^{1.6}$, and

hysteresis loss is thus given by,

$P_h \approx W_h f \approx \eta f \beta_{max}^{1.6}$

where, f is the frequency, η is the hysteresis coefficient and β_{max} is the maximum flux density, the empirical exponent of which varies from about 1.4 to 1.8 but is often given as 1.6 for iron.

Eddy current losses:

Ferromagnetic materials are also good conductors and a core made from such a material also constitutes a single short-circuited turn throughout its entire length. Eddy currents therefore circulate within the core in a plane normal to the flux, and are responsible for resistive heating of the core material. The eddy current loss is a complex function of the square of supply frequency and inverse square of the material thickness. Eddy current losses can be reduced by making the core of a stack of plates electrically insulated from each other, rather than a solid block; all transformers operating at low frequencies use laminated or similar cores.

Magnetostriction related transformer hum:

Magnetic flux in a ferromagnetic material, such as the core, causes it to physically expand and contract slightly with each cycle of the magnetic field, an effect known as magnetostriction, the frictional energy of which produces an audible noise known as mains hum or transformer hum. This transformer hum is especially objectionable in transformers supplied at power frequencies and in high-frequency flyback transformers associated with television CRTs.

Stray losses:

Leakage inductance is by itself largely lossless, since energy supplied to its magnetic fields is returned to the supply with the next half-cycle. However, any leakage flux that intercepts nearby conductive materials such as the transformer's support structure will give rise to eddy currents and be converted to heat. There are also radiative losses due to the oscillating magnetic field but these are usually small.

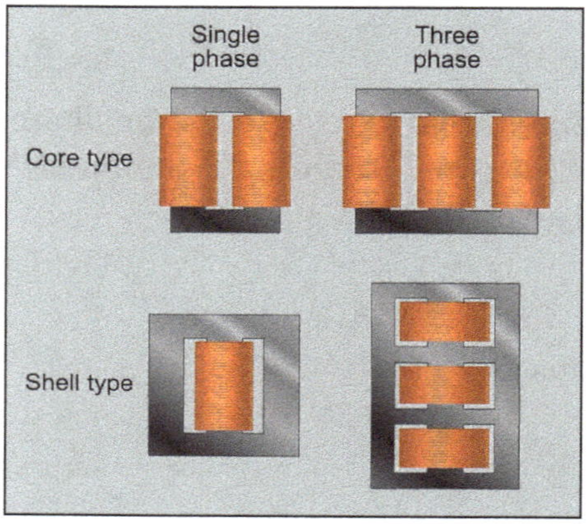

Core form = core type; shell form = shell type.

Mechanical vibration and audible noise transmission:

In addition to magnetostriction, the alternating magnetic field causes fluctuating forces between the primary and secondary windings. This energy incites vibration transmission in interconnected metalwork, thus amplifying audible transformer hum.

Core form and Shell form Transformers

Closed-core transformers are constructed in 'core form' or 'shell form'. When windings surround the core, the transformer is core form; when windings are surrounded by the core, the transformer is shell form. Shell form design may be more prevalent than core form design for distribution transformer applications due to the relative ease in stacking the core around winding coils. Core form design tends to, as a general rule, be more economical, and therefore more prevalent, than shell form design for high voltage power transformer applications at the lower end of their voltage and power rating ranges (less than or equal to, nominally, 230 kV or 75 MVA). At higher voltage and power ratings, shell form transformers tend to be more prevalent. Shell form design tends to be preferred for extra-high voltage and higher MVA applications because, though more labor-intensive to manufacture, shell form transformers are characterized as having inherently better kVA-to-weight ratio, better short-circuit strength characteristics and higher immunity to transit damage.

Cores

Laminated Steel Cores

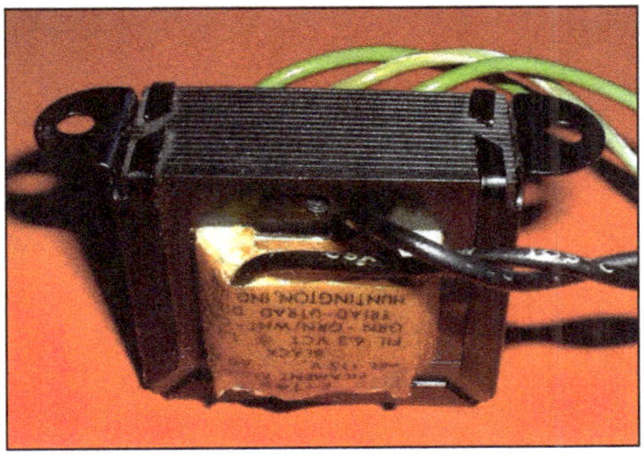

Laminated core transformer showing edge of laminations at top of photo.

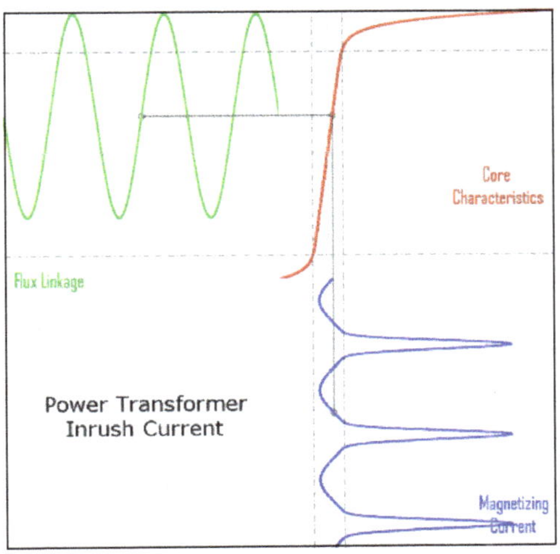

Power transformer inrush current caused by residual flux at switching instant; flux (green), iron core's magnetic characteristics (red) and magnetizing current (blue).

Transformers for use at power or audio frequencies typically have cores made of high permeability silicon steel. The steel has a permeability many times that of free space and the core thus serves to greatly reduce the magnetizing current and confine the flux to a path which closely couples the windings. Early transformer developers soon realized that cores constructed from solid iron resulted in prohibitive eddy current losses, and their designs mitigated this effect with cores consisting of bundles of insulated iron wires. Later designs constructed the core by stacking layers of thin steel laminations, a principle that has remained in use. Each lamination is insulated from its neighbors by a thin non-conducting layer of insulation. The transformer universal EMF equation implies an acceptably large core cross-sectional area in order to avoid saturation.

The effect of laminations is to confine eddy currents to highly elliptical paths that enclose little flux, and so reduce their magnitude. Thinner laminations reduce losses, but are more laborious and expensive to construct. Thin laminations are generally used on high-frequency transformers, with some of very thin steel laminations able to operate up to 10 kHz.

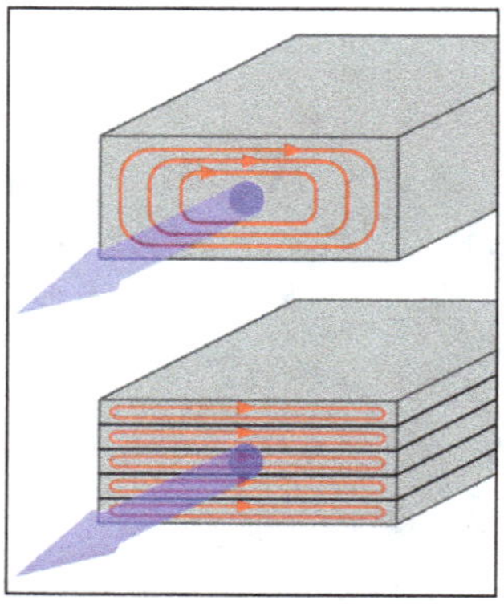

Laminating the core greatly reduces eddy-current losses.

One common design of laminated core is made from interleaved stacks of E-shaped steel sheets capped with I-shaped pieces, leading to its name of 'E-I transformer'. Such a design tends to exhibit more losses, but is very economical to manufacture. The cut-core or C-core type is made by winding a steel strip around a rectangular form and then bonding the layers together. It is then cut in two, forming two C shapes, and the core assembled by binding the two C halves together with a steel strap. They have the advantage that the flux is always oriented parallel to the metal grains, reducing reluctance.

A steel core's remanence means that it retains a static magnetic field when power is removed. When power is then reapplied, the residual field will cause a high inrush current until the effect of the remaining magnetism is reduced, usually after a few cycles of the applied AC waveform. Overcurrent protection devices such as fuses must be selected to allow this harmless inrush to pass. On transformers connected to long, overhead power transmission lines, induced currents due to geomagnetic disturbances during solar storms can cause saturation of the core and operation of transformer protection devices.

Distribution transformers can achieve low no-load losses by using cores made with low-loss high-permeability silicon steel or amorphous (non-crystalline) metal alloy. The higher initial cost of the core material is offset over the life of the transformer by its lower losses at light load.

Solid Cores

Powdered iron cores are used in circuits such as switch-mode power supplies that operate above mains frequencies and up to a few tens of kilohertz. These materials combine high magnetic permeability with high bulk electrical resistivity. For frequencies extending beyond the VHF band, cores made from non-conductive magnetic ceramic materials called ferrites are common. Some radio-frequency transformers also have movable cores (sometimes called 'slugs') which allow adjustment of the coupling coefficient (and bandwidth) of tuned radio-frequency circuits.

Toroidal Cores

Small toroidal core transformer.

Toroidal transformers are built around a ring-shaped core, which, depending on operating frequency, is made from a long strip of silicon steel or permalloy wound into a coil, powdered iron, or ferrite. A strip construction ensures that the grain boundaries are optimally aligned, improving the transformer's efficiency by reducing the core's reluctance. The closed ring shape eliminates air gaps inherent in the construction of an E-I core. The cross-section of the ring is usually square or rectangular, but more expensive cores with circular cross-sections are also available. The primary and secondary coils are often wound concentrically to cover the entire surface of the core. This minimizes the length of wire needed and provides screening to minimize the core's magnetic field from generating electromagnetic interference.

Toroidal transformers are more efficient than the cheaper laminated E-I types for a similar power level. Other advantages compared to E-I types, include smaller size (about half), lower weight (about half), less mechanical hum (making them superior in audio amplifiers), lower exterior magnetic field (about one tenth), low off-load losses (making them more efficient in standby circuits), single-bolt mounting, and greater choice of shapes. The main disadvantages are higher cost and limited power capacity. Because of the lack of a residual gap in the magnetic path, toroidal transformers also tend to exhibit higher inrush current, compared to laminated E-I types.

Ferrite toroidal cores are used at higher frequencies, typically between a few tens of kilohertz to hundreds of megahertz, to reduce losses, physical size, and weight of inductive components. A drawback of toroidal transformer construction is the higher labor cost of winding. This is because it is necessary to pass the entire length of a coil winding through the core aperture each time a single turn is added to the coil. As a consequence, toroidal transformers rated more than a few kVA are uncommon. Relatively few toroids are offered with power ratings above 10 kVA, and practically none above 25 kVA. Small distribution transformers may achieve some of the benefits of a toroidal core by splitting it and forcing it open, then inserting a bobbin containing primary and secondary windings.

Air Cores

A physical core is not an absolute requisite and a functioning transformer can be produced simply by placing the windings near each other, an arrangement termed an "air-core" transformer. The air which comprises the magnetic circuit is essentially lossless, and so an air-core transformer eliminates loss due to hysteresis in the core material. The leakage inductance is inevitably high, resulting in very poor regulation, and so such designs are unsuitable for use in power distribution. They have however very high bandwidth, and are frequently employed in radio-frequency applications, for which a satisfactory coupling coefficient is maintained by carefully overlapping the primary and secondary windings. They're also used for resonant transformers such as Tesla coils where they can achieve reasonably low loss in spite of the high leakage inductance.

Windings

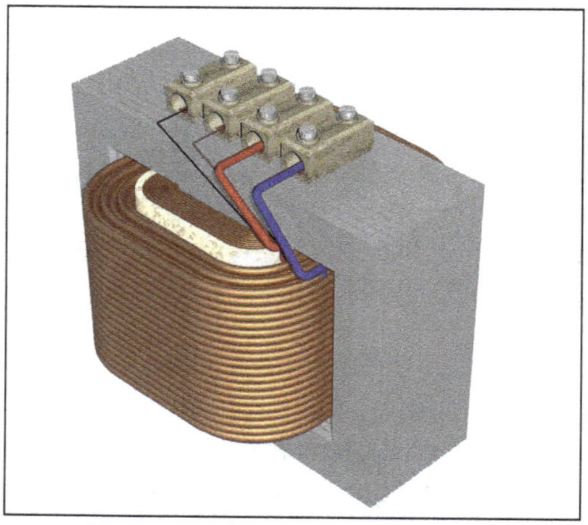

Windings are usually arranged concentrically to minimize flux leakage.

High-frequency transformers operating in the tens to hundreds of kilohertz often have windings made of braided Litz wire to minimize the skin-effect and proximity effect losses. Large power transformers use multiple-stranded conductors as well, since even at low power frequencies non-uniform distribution of current would otherwise exist in high-current windings. Each strand is individually insulated, and the strands are arranged so that at certain points in the winding, or throughout the whole winding, each portion occupies different relative positions in the complete conductor. The transposition equalizes the current flowing in each strand of the conductor, and reduces eddy current losses in the winding itself. The stranded conductor is also more flexible than a solid conductor of similar size, aiding manufacture.

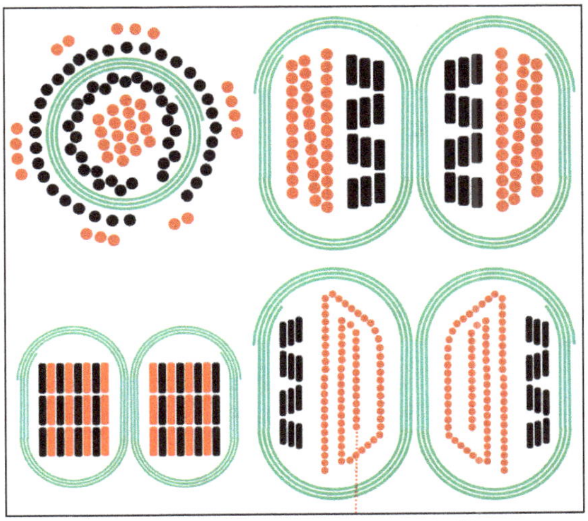

Cut view through transformer windings.

The windings of signal transformers minimize leakage inductance and stray capacitance to improve high-frequency response. Coils are split into sections, and those sections interleaved between the sections of the other winding.

Power-frequency transformers may have taps at intermediate points on the winding, usually on the higher voltage winding side, for voltage adjustment. Taps may be manually reconnected, or a manual or automatic switch may be provided for changing taps. Automatic on-load tap changers are used in electric power transmission or distribution, on equipment such as arc furnace transformers, or for automatic voltage regulators for sensitive loads. Audio-frequency transformers, used for the distribution of audio to public address loudspeakers, have taps to allow adjustment of impedance to each speaker. A center-tapped transformer is often used in the output stage of an audio power amplifier in a push-pull circuit. Modulation transformers in AM transmitters are very similar.

Dry-type transformer winding insulation systems can be either of standard open-wound 'dip-and-bake' construction or of higher quality designs that include vacuum pressure impregnation (VPI), vacuum pressure encapsulation (VPE), and cast coil encapsulation processes. In the VPI process, a combination of heat, vacuum and pressure is used to thoroughly seal, bind, and eliminate entrained air voids in the winding polyester resin insulation coat layer, thus increasing resistance to corona. VPE windings are similar to VPI windings but provide more protection against environmental effects, such as from water, dirt or corrosive ambients, by multiple dips including typically in terms of final epoxy coat.

Regarding image at top captioned, Cut view of transformer windings:

> The conducting material used for the windings depends upon the application, but in all cases the individual turns must be electrically insulated from each other to ensure that the current travels throughout every turn. For small power and signal transformers, in which currents are low and the potential difference between adjacent turns is small, the coils are often wound from enamelled magnet wire, such as Formvar wire. Larger power transformers operating at high voltages may be wound with copper rectangular strip conductors insulated by oil-impregnated paper and blocks of pressboard.
>
> Legend:
>
> White: Air, liquid or other insulating medium in conjunction with varnish, paper or other coil insulation.
>
> Green spiral: Grain oriented silicon steel.
>
> Black: Primary winding (Aluminum or copper).
>
> Red: Secondary winding (Aluminum or copper).

Cooling

Cutaway view of liquid-immersed construction transformer. The conservator (reservoir) at top provides liquid-to-atmosphere isolation as coolant level and temperature changes. The walls and fins provide required heat dissipation balance.

To place the cooling problem in perspective, the accepted rule of thumb is that the life expectancy of insulation in all electrics, including all transformers, is halved for about every 7 °C to 10 °C increase in operating temperature, this life expectancy halving rule holding more narrowly when the increase is between about 7 °C to 8 °C in the case of transformer winding cellulose insulation.

Small dry-type and liquid-immersed transformers are often self-cooled by natural convection and radiation heat dissipation. As power ratings increase, transformers are often cooled by forced-air cooling, forced-oil cooling, water-cooling, or combinations of these. Large transformers are filled with transformer oil that both cools and insulates the windings. Transformer oil is a highly refined mineral oil that cools the windings and insulation by circulating within the transformer tank. The mineral oil and paper insulation system has been extensively studied and used for more than 100 years. It is estimated that 50% of power transformers will survive 50 years of use, that the average age of failure of power transformers is about 10 to 15 years, and that about 30% of power transformer failures are due to insulation and overloading failures. Prolonged operation at elevated temperature degrades insulating properties of winding insulation and dielectric coolant, which not only shortens transformer life but can ultimately lead to catastrophic transformer failure. With a great body of empirical study as a guide, transformer oil testing including dissolved gas analysis provides valuable maintenance information. This underlines the need to monitor, model, forecast and manage oil and winding conductor insulation temperature conditions under varying, possibly difficult, power loading conditions.

Building regulations in many jurisdictions require indoor liquid-filled transformers to either use dielectric fluids that are less flammable than oil, or be installed in fire-resistant rooms. Air-cooled dry transformers can be more economical where they eliminate the cost of a fire-resistant transformer room.

The tank of liquid filled transformers often has radiators through which the liquid coolant circulates by natural convection or fins. Some large transformers employ electric fans for forced-air cooling, pumps for forced-liquid cooling, or have heat exchangers for water-cooling. An oil-immersed transformer may be equipped with a Buchholz relay, which, depending on severity of gas accumulation due to internal arcing, is used to either alarm or de-energize the transformer. Oil-immersed transformer installations usually include fire protection measures such as walls, oil containment, and fire-suppression sprinkler systems.

Polychlorinated biphenyls have properties that once favored their use as a dielectric coolant, though concerns over their environmental persistence led to a widespread ban on their use. Today, non-toxic, stable silicone-based oils, or fluorinated hydrocarbons may be used where the expense of a fire-resistant liquid offsets additional building cost for a transformer vault. PCBs for new equipment were banned in 1981 and in 2000 for use in existing equipment in United Kingdom Legislation enacted in Canada between 1977 and 1985 essentially bans PCB use in transformers manufactured in or imported into the country after 1980, the maximum allowable level of PCB contamination in existing mineral oil transformers being 50 ppm.

Some transformers, instead of being liquid-filled, have their windings enclosed in sealed, pressurized tanks and cooled by nitrogen or sulfur hexafluoride gas.

Experimental power transformers in the 500-to-1,000 kVA range have been built with liquid nitrogen or helium cooled superconducting windings, which eliminates winding losses without affecting core losses.

Insulation Drying

Construction of oil-filled transformers requires that the insulation covering the windings be thoroughly dried of residual moisture before the oil is introduced. Drying is carried out at the factory, and may also be required as a field service. Drying may be done by circulating hot air around the core, or by vapor-phase drying (VPD) where an evaporated solvent transfers heat by condensation on the coil and core.

For small transformers, resistance heating by injection of current into the windings is used. The heating can be controlled very well, and it is energy efficient. The method is called low-frequency heating (LFH) since the current used is at a much lower frequency than that of the power grid, which is normally 50 or 60 Hz. A lower frequency reduces the effect of inductance, so the voltage required can be reduced. The LFH drying method is also used for service of older transformers.

Bushings

Larger transformers are provided with high-voltage insulated bushings made of polymers or porcelain. A large bushing can be a complex structure since it must provide careful control of the electric field gradient without letting the transformer leak oil.

Classification Parameters

Transformers can be classified in many ways, such as the following:

- *Power capacity*: From a fraction of a volt-ampere (VA) to over a thousand MVA.
- *Duty of a transformer*: Continuous, short-time, intermittent, periodic, varying.
- *Frequency range*: Power-frequency, audio-frequency, or radio-frequency.
- *Voltage class*: From a few volts to hundreds of kilovolts.
- *Cooling type*: Dry and liquid-immersed – self-cooled, forced air-cooled; liquid-immersed – forced oil-cooled, water-cooled.
- *Circuit application*: Such as power supply, impedance matching, output voltage and current stabilizer or circuit isolation.
- *Utilization*: Pulse, power, distribution, rectifier, arc furnace, amplifier output, etc.
- *Basic magnetic form*: Core form, shell form, concentric, sandwich.

- *Constant-potential transformer descriptor*: Step-up, step-down, isolation.
- *General winding configuration*: By EIC vector group – various possible two-winding combinations of the phase designations delta, wye or star, and zigzag or interconnected star; other – autotransformer, Scott-T, zigzag grounding transformer winding.
- *Rectifier phase-shift winding configuration*: 2-winding, 6-pulse; 3-winding, 12-pulse; . . . n-winding, [n-1]*6-pulse; polygon; etc.

Types

Various specific electrical application designs require a variety of transformer types. Although they all share the basic characteristic transformer principles, they are customize in construction or electrical properties for certain installation requirements or circuit conditions:

- *Autotransformer*: Transformer in which part of the winding is common to both primary and secondary circuits, leading to increased efficiency, smaller size, and a higher degree of voltage regulation.
- *Capacitor voltage transformer*: Transformer in which capacitor divider is used to reduce high voltage before application to the primary winding.
- *Distribution transformer, power transformer*: International standards make a distinction in terms of distribution transformers being used to distribute energy from transmission lines and networks for local consumption and power transformers being used to transfer electric energy between the generator and distribution primary circuits.
- *Phase angle regulating transformer*: A specialised transformer used to control the flow of real power on three-phase electricity transmission networks.
- *Scott-T transformer*: Transformer used for phase transformation from three-phase to two-phase and vice versa.
- *Polyphase transformer*: Any transformer with more than one phase.
- *Grounding transformer*: Transformer used for grounding three-phase circuits to create a neutral in a three wire system, using a wye-delta transformer, or more commonly, a zigzag grounding winding.
- *Leakage transformer*: Transformer that has loosely coupled windings.
- *Resonant transformer*: Transformer that uses resonance to generate a high secondary voltage.

- *Audio transformer*: Transformer used in audio equipment.
- *Output transformer*: Transformer used to match the output of a valve amplifier to its load.
- *Instrument transformer*: Potential or current transformer used to accurately and safely represent voltage, current or phase position of high voltage or high power circuits.
- *Pulse transformer*: Specialized small-signal transformer used to transmit digital signaling while providing electrical isolation, commonly used in Ethernet computer networks as 10BASE-T, 100BASE-T and 1000BASE-T.

An electrical substation in Melbourne, Australia showing three of five 220 kV – 66 kV transformers, each with a capacity of 150 MVA.

Applications

Transformer at the Limestone Generating Station in Manitoba, Canada.

Transformers are used to increase (or step-up) voltage before transmitting electrical energy over long distances through wires. Wires have resistance which loses energy through joule heating at a rate corresponding to square of the current. By transforming power to a higher voltage transformers enable economical transmission of power and distribution. Consequently, transformers have shaped the electricity supply industry, permitting generation to be located remotely from points of demand. All but a tiny fraction of the world's electrical power has passed through a series of transformers by the time it reaches the consumer.

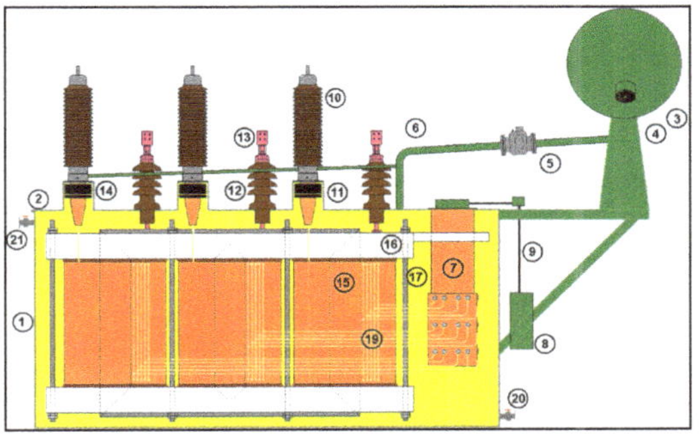

Schematic of a large oil filled power transformer.

Since the high voltages carried in the wires are significantly greater than what is needed in-home, transformers are also used extensively in electronic products to decrease (or step-down) the supply voltage to a level suitable for the low voltage circuits they contain. The transformer also electrically isolates the end user from contact with the supply voltage.

Signal and audio transformers are used to couple stages of amplifiers and to match devices such as microphones and record players to the input of amplifiers. Audio transformers allowed telephone circuits to carry on a two-way conversation over a single pair of wires. A balun transformer converts a signal that is referenced to ground to a signal that has balanced voltages to ground, such as between external cables and internal circuits. Transformers made to medical grade standards isolate the users from the direct current. These are found commonly used in conjunction with hospital beds, dentist chairs, and other medical lab equipment.

History

Discovery of Induction

Faraday's experiment with induction between coils of wire.

Electromagnetic induction, the principle of the operation of the transformer, was discovered independently by Michael Faraday in 1831, Joseph Henry in 1832, and others. The relationship between EMF and magnetic flux is an equation now known as Faraday's law of induction:

$$|\mathcal{E}| = \left|\frac{d\Phi_B}{dt}\right|.$$

where $|\mathcal{E}|$ is the magnitude of the EMF in Volts and Φ_B is the magnetic flux through the circuit in webers.

Faraday performed early experiments on induction between coils of wire, including winding a pair of coils around an iron ring, thus creating the first toroidal closed-core transformer. However he only applied individual pulses of current to his transformer, and never discovered the relation between the turns ratio and EMF in the windings.

Induction coil, 1900, Bremerhaven, Germany.

Induction Coils

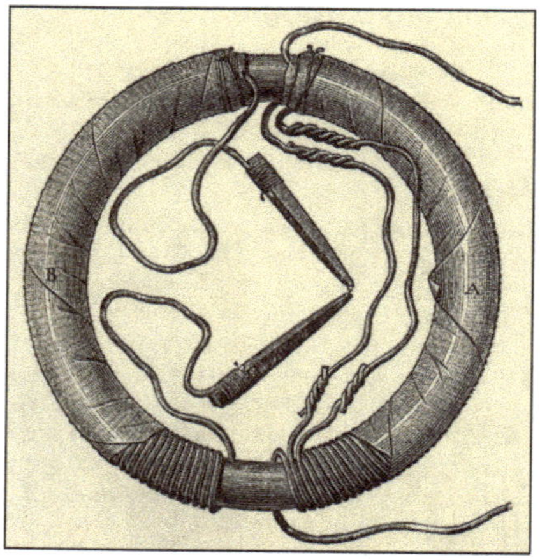

Faraday's ring transformer.

The first type of transformer to see wide use was the induction coil, invented by Rev. Nicholas Callan of Maynooth College, Ireland in 1836. He was one of the first researchers to realize the more turns the secondary winding has in relation to the primary winding, the larger the induced secondary EMF will be. Induction coils evolved from scientists' and inventors' efforts to get higher voltages from batteries. Since batteries produce direct current (DC) rather than AC, induction coils relied upon vibrating electrical contacts that regularly interrupted the current in the primary to create the flux changes necessary for induction. Between the 1830s and the 1870s, efforts to build better induction coils, mostly by trial and error, slowly revealed the basic principles of transformers.

First Alternating Current Transformers

By the 1870s, efficient generators producing alternating current (AC) were available, and it was found AC could power an induction coil directly, without an interrupter.

In 1876, Russian engineer Pavel Yablochkov invented a lighting system based on a set of induction coils where the primary windings were connected to a source of AC. The secondary windings could be connected to several 'electric candles' (arc lamps) of his own design. The coils Yablochkov employed functioned essentially as transformers.

In 1878, the Ganz factory, Budapest, Hungary, began equipment for electric lighting and, by 1883, had installed over fifty systems in Austria-Hungary. Their AC systems used arc and incandescent lamps, generators, and other equipment.

Lucien Gaulard and John Dixon Gibbs first exhibited a device with an open iron core called a 'secondary generator' in London in 1882, then sold the idea to the Westinghouse company in the United States. They also exhibited the invention in Turin, Italy in 1884, where it was adopted for an electric lighting system.

Early Series Circuit Transformer Distribution

Induction coils with open magnetic circuits are inefficient at transferring power to loads. Until about 1880, the paradigm for AC power transmission from a high voltage supply to a low voltage load was a series circuit. Open-core transformers with a ratio near 1:1 were connected with their primaries in series to allow use of a high voltage for transmission while presenting a low voltage to the lamps. The inherent flaw in this method was that turning off a single lamp (or other electric device) affected the voltage supplied to all others on the same circuit. Many adjustable transformer designs were introduced to compensate for this problematic characteristic of the series circuit, including those employing methods of adjusting the core or bypassing the magnetic flux around part of a coil. Efficient, practical transformer designs did not appear until the 1880s, but within a decade, the transformer would be instrumental in the War of Currents, and in seeing AC distribution systems triumph over their DC counterparts, a position in which they have remained dominant ever since.

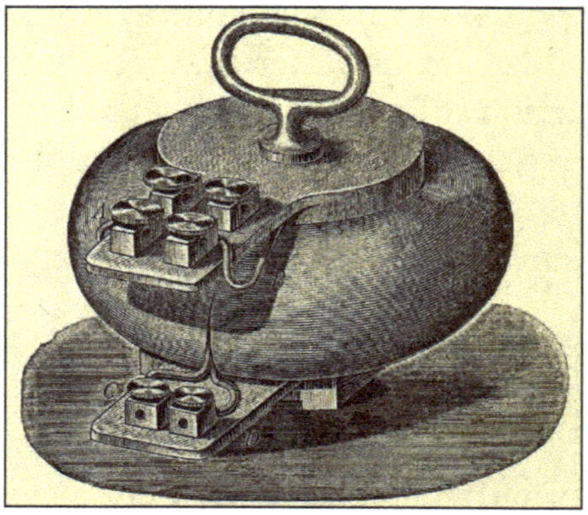

Shell form transformer.
Sketch used by Uppenborn to describe
ZBD engineers' 1885 patents and earliest articles.

Core form, front; shell form, back. Earliest specimens of ZBD-designed high-efficiency constant-potential transformers manufactured at the Ganz factory in 1885.

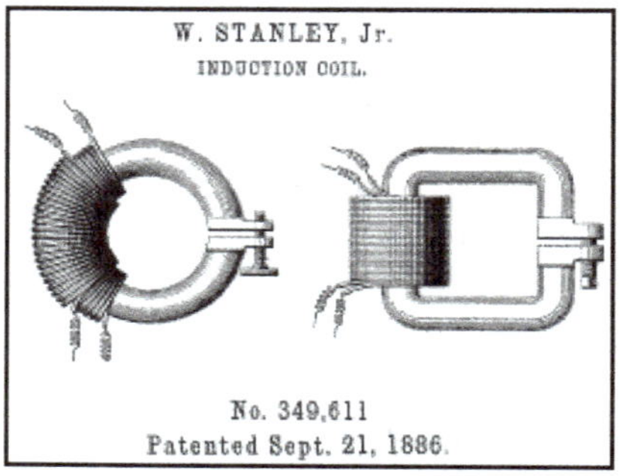

Stanley's 1886 design for adjustable gap open-core induction coils.

Closed-core Transformers and Parallel Power Distribution

In the autumn of 1884, Károly Zipernowsky, Ottó Bláthy and Miksa Déri (ZBD), three engineers associated with the Ganz factory, had determined that open-core devices were impracticable, as they were incapable of reliably regulating voltage. In their joint 1885 patent applications for novel transformers (later called ZBD transformers), they described two designs with closed magnetic circuits where copper windings were either a) wound around iron wire ring core or b) surrounded by iron wire core. The two designs were the first application of the two basic transformer constructions in common use to this day, which can as a class all be termed as either core form or shell form (or alternatively, core type or shell type), as in a) or b), respectively. The Ganz factory had also in the autumn of 1884 made delivery of the world's first five high-efficiency AC transformers, the first of these units having been shipped on September 16, 1884. This first unit had been manufactured to the following specifications: 1,400 W, 40 Hz, 120:72 V, 11.6:19.4 A, ratio 1.67:1, one-phase, shell form.

In both designs, the magnetic flux linking the primary and secondary windings traveled almost entirely within the confines of the iron core, with no intentional path through air. The new transformers were 3.4 times more efficient than the open-core bipolar devices of Gaulard and Gibbs. The ZBD patents included two other major interrelated innovations: one concerning the use of parallel connected, instead of series connected, utilization loads, the other concerning the ability to have high turns ratio transformers such that the supply network voltage could be much higher (initially 1,400 to 2,000 V) than the voltage of utilization loads (100 V initially preferred). When employed in parallel connected electric distribution systems, closed-core transformers finally made it technically and economically feasible to provide electric power for lighting in homes, businesses and public spaces. Bláthy had suggested the use of closed cores, Zipernowsky had suggested the use of parallel shunt connections, and Déri had performed the experiments.

Transformers today are designed on the principles discovered by the three engineers. They also popularized the word 'transformer' to describe a device for altering the EMF of an electric current, although the term had already been in use by 1882. In 1886, the ZBD engineers designed, and the Ganz factory supplied electrical equipment for, the world's first power station that used AC generators to power a parallel connected common electrical network, the steam-powered Rome-Cerchi power plant.

Although George Westinghouse had bought Gaulard and Gibbs' patents in 1885, the Edison Electric Light Company held an option on the US rights for the ZBD transformers, requiring Westinghouse to pursue alternative designs on the same principles. He assigned to William Stanley the task of developing a device for commercial use in United States. Stanley's first patented design was for induction coils with single cores of soft iron and adjustable gaps to regulate the EMF present in the secondary winding. This design was first used commercially in the US in 1886 but Westinghouse was intent on improving the Stanley design to make it (unlike the ZBD type) easy and cheap to produce.

Westinghouse, Stanley and associates soon developed an easier to manufacture core, consisting of a stack of thin 'Eshaped' iron plates, insulated by thin sheets of paper or other insulating material. Prewound copper coils could then be slid into place, and straight iron plates laid in to create a closed magnetic circuit. Westinghouse applied for a patent for the new low-cost design in December 1886; it was granted in July 1887.

Other Early Transformer Designs

In 1889, Russian-born engineer Mikhail Dolivo-Dobrovolsky developed the first three-phase transformer at the Allgemeine Elektricitäts-Gesellschaft ('General Electricity Company') in Germany.

In 1891, Nikola Tesla invented the Tesla coil, an air-cored, dual-tuned resonant transformer for producing very high voltages at high frequency.

Audio frequency transformers ('repeating coils') were used by early experimenters in the development of the telephone.

Transformer Types

An electric arc furnace the transformer has heavy copper bus for the low voltage winding, which can be rated for tens of thousands of amperes. They are immersed in oil for cooling and insulation, and are designed to survive frequent short circuits.

A variety of types of electrical transformer are made for different purposes. Despite their design differences, the various types employ the same basic principle as discovered in 1831 by Michael Faraday, and share several key functional parts.

Power Transformers

Laminated Core

Laminated core transformer.

This is the most common type of transformer, widely used in electric power transmission and appliances to convert mains voltage to low voltage to power electronic devices. They are available in power ratings ranging from mW to MW. The insulated laminations minimizes eddy current losses in the iron core.

Small appliance and electronic transformers may use a split bobbin, giving a high level of insulation between the windings. The rectangular cores are made up of stampings, often in E-I shape pairs, but other shapes are sometimes used. Shields between primary and secondary may be fitted to reduce EMI (electromagnetic interference), or a screen winding is occasionally used.

Small appliance and electronics transformers may have a thermal cut-out built into the winding, to shut-off power at high temperatures to prevent further overheating.

Toroidal

Doughnut shaped toroidal transformers save space compared to E-I cores, and sometimes to reduce external magnetic field. These use a ring shaped core, copper windings wrapped round this ring (and thus threaded through the ring during winding), and tape for insulation.

Toroidal transformers have a lower external magnetic field compared to rectangular transformers, and can be smaller for a given power rating. However, they cost more to make, as winding requires more complex and slower equipment.

They can be mounted by a bolt through the center, using washers and rubber pads or by potting in resin.

Autotransformer

An autotransformer has one winding that is tapped at some point along the winding. Voltage is applied across a portion of the winding, and a higher (or lower) voltage is produced across another portion of the same winding. The equivalent power rating of the autotransfomer is lower than the actual load power rating. It is calculated by: load VA × (|Vin – Vout|)/Vin. For example, an auto transformer that adapts a 1000 VA load rated at 120 Volts to a 240 Volt supply has an equivalent rating of at least: 1,000VA × (240V – 120V) / 240V = 500VA. However, the actual rating (shown on the tally plate) must be at least 1000 VA.

For voltage ratios that don't exceed about 3:1, an autotransformer is cheaper, lighter, smaller, and more efficient than an isolating (two-winding) transformer of the same rating. Large three-phase autotransformers are used in electric power distribution systems, for example, to interconnect 33 kV and 66 kV sub-transmission networks.

Variable Autotransformer

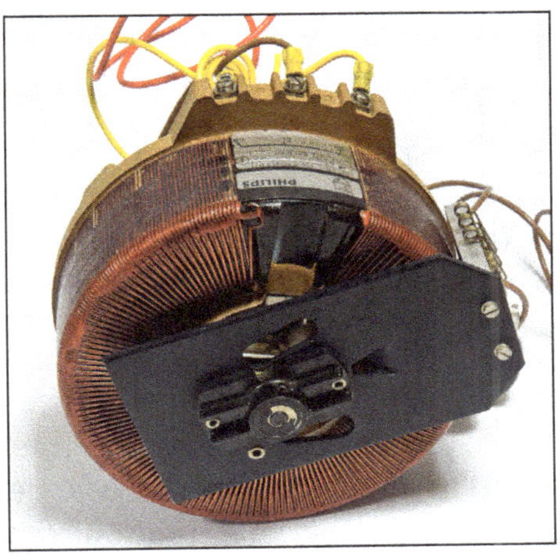

Variable autotransformer.

By exposing part of the winding coils of an autotransformer, and making the secondary connection through a sliding carbon brush, an autotransformer with a near-continuously variable turns ratio can be obtained, allowing for wide voltage adjustment in very small increments.

Induction Regulator

The induction regulator is similar in design to a wound-rotor induction motor but it is essentially a transformer whose output voltage is varied by rotating its secondary relative to the primary—i.e., rotating the angular position of the rotor. It can be seen as a power transformer exploiting rotating magnetic fields. The major advantage of the induction regulator is that unlike variacs, they are practical for transformers over 5 kVA. Hence, such regulators find widespread use in high-voltage laboratories.

Polyphase Transformer

A high-voltage transformer being dismantled.

Three-phase transformers 380kV/110kV and 110kV/20kV.

For polyphase systems, multiple single-phase transformers can be used, or all phases can be connected to a single polyphase transformer. For a three phase transformer, the three primary windings are connected together and the three secondary windings are connected together. Examples of connections are wye-delta, delta-wye, delta-delta and wye-wye. A vector group indicates the configuration of the windings and the phase angle difference between them. If a winding is connected to earth (grounded), the earth connection point is usually the center point of a wye winding. If the secondary is a delta winding, the ground may be connected to a center tap on one winding (high leg delta) or one phase may be grounded (corner grounded delta). A special purpose polyphase transformer is the zigzag transformer. There are many possible configurations that may involve more or fewer than six windings and various tap connections.

Grounding Transformer

Grounding transformers let three wire (delta) polyphase system supplies accommodate phase to neutral loads by providing a return path for current to a neutral. Grounding transformers most commonly incorporate a single winding transformer with a zigzag winding configuration but may also be created with a wye-delta isolated winding transformer connection.

Phase-shifting Transformer

This is a specialized type of transformer which can be configured to adjust the phase relationship between input and output. This allows power flow in an electric grid to be controlled, e.g. to steer power flows away from a shorter (but overloaded) link to a longer path with excess capacity.

Variable-frequency Transformer

A *variable-frequency transformer* is a specialized three-phase power transformer which allows the phase relationship between the input and output windings to be continuously adjusted by rotating one half. They are used to interconnect electrical grids with the same nominal frequency but without synchronous phase coordination.

Leakage or Stray Field Transformers

Leakage transformer.

A leakage transformer, also called a stray-field transformer, has a significantly higher leakage inductance than other transformers, sometimes increased by a magnetic bypass or shunt in its core between primary and secondary, which is sometimes adjustable with a set screw. This provides a transformer with an inherent current limitation due to the loose coupling between its primary and the secondary windings. In this case, it is short-circuit inductance which is actually acting as a current limiting parameter. The output and input currents are low enough to prevent thermal overload under all load conditions—even if the secondary is shorted.

Uses

Leakage transformers are used for arc welding and high voltage discharge lamps (neon lights and cold cathode fluorescent lamps, which are series connected up to 7.5 kV AC). It acts then both as a voltage transformer and as a magnetic ballast.

Other applications are short-circuit-proof extra-low voltage transformers for toys or doorbell installations.

Resonant Transformer

A resonant transformer is a transformer in which one or both windings has a capacitor across it and functions as a tuned circuit. Used at radio frequencies, resonant transformers can function as high Q_factor bandpass filters. The transformer windings have either air or ferrite cores and the bandwidth can be adjusted by varying the coupling (mutual inductance). One common form is the IF (intermediate frequency) transformer, used in superheterodyne radio receivers. They are also used in radio transmitters.

Resonant transformers are also used in electronic ballasts for gas discharge lamps, and high voltage power supplies. They are also used in some types of switching power supplies. Here the short-circuit inductance value is an important parameter that determines the resonance frequency of the resonant transformer. Often only secondary winding has a resonant capacitor (or stray capacitance) and acts as a serial resonant tank circuit. When the short-circuit inductance of the secondary side of the transformer is L_{sc} and the resonant capacitor (or stray capacitance) of the secondary side is C_r, The resonance frequency ω_s of 1' is as follows:

$$\omega_s = \frac{1}{\sqrt{L_{sc}C_r}} = \frac{1}{\sqrt{(1-k^2)L_sC_r}}$$

The transformer is driven by a pulse or square wave for efficiency, generated by an electronic oscillator circuit. Each pulse serves to drive resonant sinusoidal oscillations in the tuned winding, and due to resonance a high voltage can be developed across the secondary.

Applications:

- Intermediate frequency (IF) transformer in superheterodyne radio receiver.
- Tank transformers in radio transmitters.
- Tesla coil.
- CCFL inverter.
- Oudin coil (or Oudin resonator; named after its inventor Paul Oudin).
- D'Arsonval apparatus.
- Ignition coil or induction coil used in the ignition system of a petrol engine.
- Electrical breakdown and insulation testing of high voltage equipment and cables. In the latter case, the transformer's secondary is resonated with the cable's capacitance.

Constant Voltage Transformer

By arranging particular magnetic properties of a transformer core, and installing a ferro-resonant tank circuit (a capacitor and an additional winding), a transformer can be arranged to automatically keep the secondary winding voltage relatively constant for varying primary supply without additional circuitry or manual adjustment. Ferro-resonant transformers run hotter than standard power transformers, because regulating action depends on core saturation, which reduces efficiency. The output waveform is heavily distorted unless careful measures are taken to prevent this. Saturating transformers provide a simple rugged method to stabilize an AC power supply.

Ferrite Core

Ferrite core power transformers are widely used in switched-mode power supplies (SMPSs). The powder core enables high-frequency operation, and hence much smaller size-to-power ratio than laminated-iron transformers.

Ferrite transformers are not used as power transformers at mains frequency since laminated iron cores cost less than an equivalent ferrite core.

Planar Transformer

Manufacturers etch spiral patterns on a printed circuit board to form the "windings" of a planar transformer, replacing the turns of wire used to make other types. Some planar transformers are commercially sold as discrete components. Other planar transformers are one of many components on a printed circuit board. A planar transformer can be thinner than other transformers, which is useful for low-profile applications or when several printed circuit boards are stacked. Almost all planar transformers use a ferrite planar core.

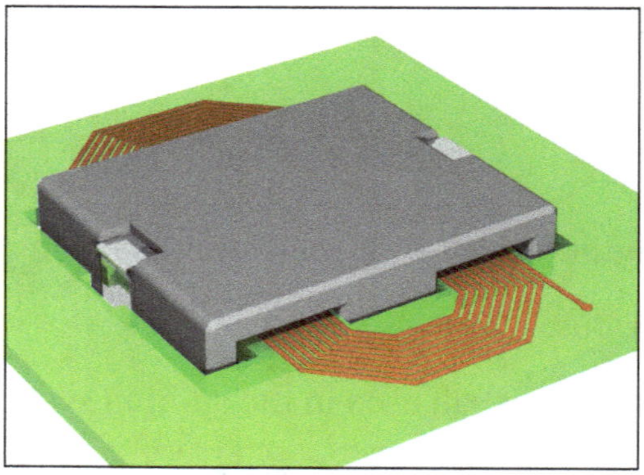

A planar transformer.

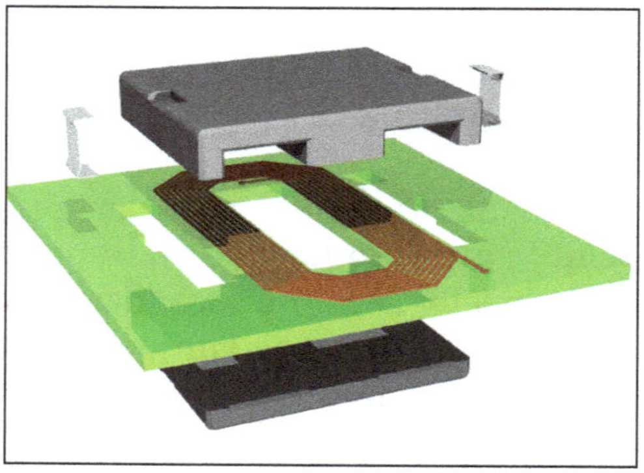

Exploded view: the spiral primary "winding" on one side of the PCB (the spiral secondary "winding" is on the other side of the PCB).

Oil Cooled Transformer

Large transformers used in power distribution or electrical substations have their core and coils immersed in oil, which cools and insulates. Oil circulates through ducts in the coil and around the coil and core assembly, moved by convection. The oil is cooled by the outside of the tank in small ratings, and by an air-cooled radiator in larger ratings. Where a higher rating is required, or where the transformer is in a building or underground, oil pumps circulate the oil, and an oil-to-water heat exchanger may also be used. Some transformers may contain PCBs where or when its use was permitted. For example, until 1979 in South Africa. substitute fire-resistant liquids such as silicone oils are now used instead.

Cast Resin Transformer

Cast-resin power transformers encase the windings in epoxy resin. These transformers simplify installation since they are dry, without cooling oil, and so require no fire-proof vault for indoor installations. The epoxy protects the windings from dust and corrosive atmospheres. However, because the molds for casting the coils are only available in fixed sizes, the design of the transformers is less flexible, which may make them more costly if customized features (voltage, turns ratio, taps) are required.

Isolating Transformer

An isolation transformer links two circuits magnetically, but provides no metallic conductive path between the circuits. An example application would be in the power supply for medical equipment, when it is necessary to prevent any leakage from the AC power system into devices connected to a patient. Special purpose isolation transformers may include shielding to prevent coupling of electromagnetic noise between circuits, or may have reinforced insulation to withstand thousands of volts of potential difference between primary and secondary circuits.

Instrument Transformer

Instrument transformers.

Instrument transformers are typically used to operate instruments from high voltage lines or high current circuits, safely isolating measurement and control circuitry from the high voltages or currents. The primary winding of the transformer is connected to the high voltage or high current circuit, and the meter or relay is connected to the secondary circuit. Instrument transformers may also be used as an isolation transformer so that secondary quantities may be used without affecting the primary circuitry.

Terminal identifications (either alphanumeric such as H_1, X_1, Y_1, etc. or a colored spot or dot impressed in the case) indicate one end of each winding, indicating the same instantaneous polarity and phase between windings. This applies to both types of instrument transformers. Correct identification of terminals and wiring is essential for proper operation of metering and protective relay instrumentation.

Current Transformer

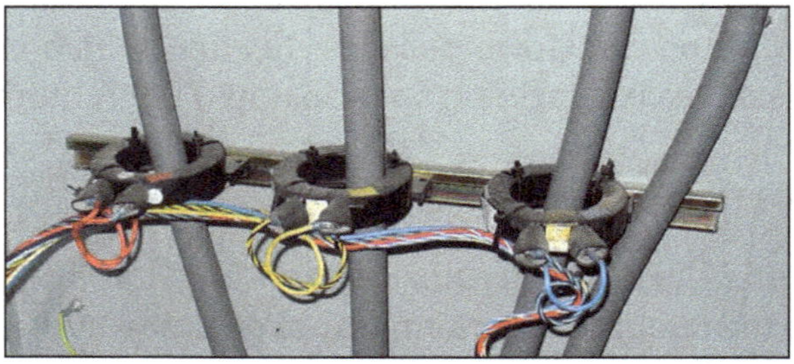

Current transformers used in metering equipment for three-phase 400 ampere electricity supply.

A current transformer (CT) is a series connected measurement device designed to provide a current in its secondary coil proportional to the current flowing in its primary. Current transformers are commonly used in metering and protective relays in the electrical power industry.

Current transformers are often constructed by passing a single primary turn (either an insulated cable or an uninsulated bus bar) through a well-insulated toroidal core wrapped with many turns of wire. The CT is typically described by its current ratio from primary to secondary. For example, a 1000:1 CT provides an output current of 1 amperes when 1000 amperes flow through the primary winding. Standard secondary current ratings are 5 amperes or 1 ampere, compatible with standard measuring instruments. The secondary winding can be single ratio or have several tap points to provide a range of ratios. Care must be taken to make sure the secondary winding is not disconnected from its low-impedance load while current flows in the primary, as this may produce a dangerously high voltage across the open secondary and may permanently affect the accuracy of the transformer.

Specially constructed wideband CTs are also used, usually with an oscilloscope, to measure high frequency waveforms or pulsed currents within pulsed power systems. One type provides a voltage output that is proportional to the measured current. Another, called a Rogowski coil, requires an external integrator in order to provide a proportional output.

A current clamp uses a current transformer with a split core that can be easily wrapped around a conductor in a circuit. This is a common method used in portable current measuring instruments but permanent installations use more economical types of current transformer.

Voltage Transformer or Potential Transformer

Voltage transformers (VT), also called potential transformers (PT), are a parallel connected type of instrument transformer, used for metering and protection in high-voltage circuits or phasor phase shift isolation. They are designed to present negligible load to the supply being measured and to have an accurate voltage ratio to enable accurate metering. A potential transformer may have several secondary windings on the same core as a primary winding, for use in different metering or protection circuits. The primary may be connected phase to ground or phase to phase. The secondary is usually grounded on one terminal.

There are three primary types of voltage transformers (VT): electromagnetic, capacitor, and optical. The electromagnetic voltage transformer is a wire-wound transformer. The capacitor voltage transformer uses a capacitance potential divider and is used at higher voltages due to a lower cost than an electromagnetic VT. An optical voltage transformer exploits the electrical properties of optical materials. Measurement of high voltages is possible by the potential transformers.

Combined Instrument Transformer

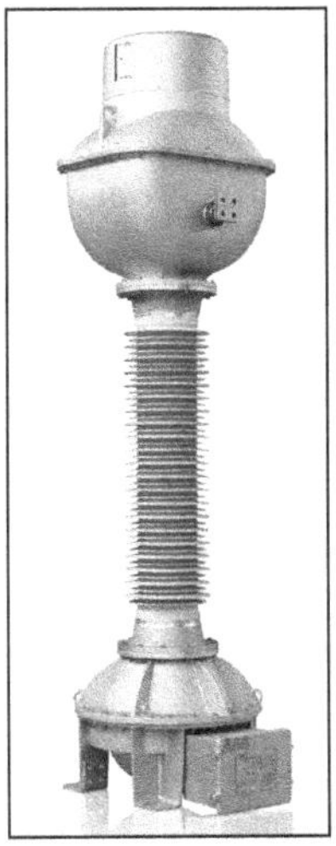

Combined instrument transformer.

A combined instrument transformer encloses a current transformer and a voltage transformer in the same transformer. There are two main combined current and voltage transformer designs: oil-paper insulated and SF_6 insulated. One advantage of applying this solution is reduced substation footprint, due to reduced number of transformers in a bay, supporting structures and connections as well as lower costs for civil works, transportation and installation.

Pulse Transformer

A pulse transformer is a transformer that is optimised for transmitting rectangular electrical pulses (that is, pulses with fast rise and fall times and a relatively constant amplitude). Small versions called *signal* types are used in digital logic and telecommunications circuits, often for matching logic drivers to transmission lines. Medium-sized *power* versions are used in power-control circuits such as camera flash controllers.

Larger *power* versions are used in the electrical power distribution industry to interface low-voltage control circuitry to the high-voltage gates of power semiconductors. Special high voltage pulse transformers are also used to generate high power pulses for radar, particle accelerators, or other high energy pulsed power applications.

Bothhand TS6121A pulse transformer.

To minimize distortion of the pulse shape, a pulse transformer needs to have low values of leakage inductance and distributed capacitance, and a high open-circuit inductance. In power-type pulse transformers, a low coupling capacitance (between the primary and secondary) is important to protect the circuitry on the primary side from high-powered transients created by the load. For the same reason, high insulation resistance and high breakdown voltage are required. A good transient response is necessary to maintain the rectangular pulse shape at the secondary, because a pulse with slow edges would create switching losses in the power semiconductors.

The product of the peak pulse voltage and the duration of the pulse (or more accurately, the voltage-time integral) is often used to characterise pulse transformers. Generally speaking, the larger this product, the larger and more expensive the transformer.

Pulse transformers by definition have a duty cycle of less than 0.5; whatever energy stored in the coil during the pulse must be "dumped" out before the pulse is fired again.

RF Transformer

There are several types of transformer used in radio frequency (RF) work. Laminated steel is not suitable for RF.

Air-core Transformer

These are used for high frequency work. The lack of a core means very low inductance. All current excites current and induces secondary voltage which is proportional to the mutual inductance. Such transformers may be nothing more than a few turns of wire soldered onto a printed circuit board.

Ferrite-core Transformer

Ferrite-core transformers are widely used in impedance matching transformers for RF, especially for baluns for TV and radio antennas. Many only have one or two turns.

Transmission-line Transformer

For radio frequency use, transformers are sometimes made from configurations of transmission line, sometimes bifilar or coaxial cable, wound around ferrite or other types of core. This style of transformer gives an extremely wide bandwidth but only a limited number of ratios (such as 1:9, 1:4 or 1:2) can be achieved with this technique.

The core material increases the inductance dramatically, thereby raising its Q factor. The cores of such transformers help improve performance at the lower frequency end of the band. RF transformers sometimes used a third coil (called a tickler winding) to inject feedback into an earlier (detector) stage in antique regenerative radio receivers.

In RF and microwave systems, a quarter-wave impedance transformer provides a way of matching impedances between circuits over a limited range of frequencies, using only a length of transmission line. The line may be coaxial cable, waveguide, stripline or microstripline.

Balun

Baluns are transformers designed specifically to connect between balanced and unbalanced circuits. These are sometimes made from configurations of transmission line and sometimes bifilar or coaxial cable and are similar to transmission line transformers in construction and operation.

IF Transformer

Ferrite-core transformers are widely used in (intermediate frequency) (IF) stages in superheterodyne radio receivers. They are mostly tuned transformers, containing a threaded ferrite slug that is screwed in or out to adjust IF tuning. The transformers are usually canned (shielded) for stability and to reduce interference.

Audio Transformer

Audio transformers are those specifically designed for use in audio circuits to carry audio signal. They can be used to block radio frequency interference or the DC component of an audio signal, to split or combine audio signals, or to provide impedance matching between high impedance and low impedance circuits, such as between a high impedance tube (valve) amplifier output and a low impedance loudspeaker, or between a high impedance instrument output and the low impedance input of a mixing console.

Audio transformers that operate with loudspeaker voltages and current are larger than those that operate at microphone or line level, which carry much less power. Bridge transformers connect 2-wire and 4-wire communication circuits.

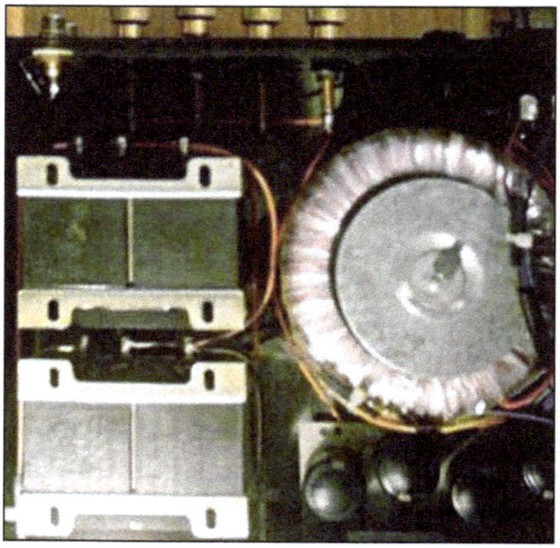

Two speaker-level audio transformers in a tube amplifier are seen on the left. The power supply toroidal transformer is on right.

Five audio transformers for various line level purposes. The two black boxes on the left contain 1:1 transformers for splitting signals, balancing unbalanced signals, or isolating two different AC ground systems to eliminate buzz and hum. The two cylindrical metal cases fit into octal sockets; each one contains a 1:1 line transformer, the first is rated at 600 ohms, the second is rated at 15,000 ohms. On the far right is a DI unit; its 12:1 transformer (with yellow insulation) changes a high impedance unbalanced input to a low impedance balanced output.

Being magnetic devices, audio transformers are susceptible to external magnetic fields such as those generated by AC current-carrying conductors. "Hum" is a term commonly used to describe unwanted signals originating from the "mains" power supply (typically 50 or 60 Hz). Audio transformers used for low-level signals, such as those from microphones, often include magnetic shielding to protect against extraneous magnetically coupled signals.

Audio transformers were originally designed to connect different telephone systems to one another while keeping their respective power supplies isolated, and are still commonly used to interconnect professional audio systems or system components, to eliminate buzz and hum. Such transformers typically have a 1:1 ratio between the primary and the secondary. These can also be used for splitting signals, balancing unbalanced signals, or feeding a balanced signal to unbalanced equipment. Transformers are also used in DI boxes to convert high-impedance instrument signals (e.g., bass guitar) to low impedance signals to enable them to connect to a microphone input on the mixing console.

A particularly critical component is the output transformer of a valve amplifier. Valve circuits for quality reproduction have long been produced with no other (inter-stage) audio transformers, but an output transformer is needed to couple the relatively high impedance (up to a few hundred ohms depending upon configuration) of the output valve(s) to the low impedance of a loudspeaker. (The valves can deliver a low current at a high voltage; the speakers require high current at low voltage.) Most solid-state power amplifiers need no output transformer at all.

Audio transformers affect the sound quality because they are non-linear. Harmonic distortion is added to the original signal, especially odd-order harmonics with an emphasis on third-order harmonics. When the incoming signal amplitude is very low there is not enough level to energize the magnetic core. When the incoming signal amplitude is very high the transformer saturates and adds ringing harmonics. Another non-linearity comes from limited frequency response. For good low-frequency response a relatively large magnetic core is required; high power handling increases the required core size. Good high-frequency response requires carefully designed and implemented windings without excessive leakage inductance or stray capacitance. All this makes for an expensive component.

Early transistor audio power amplifiers often had output transformers, but they were eliminated as advances in semiconductors allowed the design of amplifiers with sufficiently low output impedance to drive a loudspeaker directly.

Loudspeaker Transformer

Loudspeaker transformer in old radio.

In the same way that transformers create high voltage power transmission circuits that minimize transmission losses, loudspeaker transformers can power many individual loudspeakers from a single audio circuit operated at higher than normal loudspeaker voltages. This application is common in public address applications. Such circuits are commonly referred to as constant voltage speaker systems. Such systems are also known by the nominal voltage of the loudspeaker line, such as *25-*, *70-* and *100-volt* speaker systems (the voltage corresponding to the power rating of a speaker or amplifier). A transformer steps up the output of the system's amplifer to the distribution voltage. At the distant loudspeaker locations, a step-down transformer matches the speaker to the rated voltage of the line, so the speaker produces rated nominal output when the line is at nominal voltage. Loudspeaker transformers commonly have multiple primary taps to adjust the volume at each speaker in steps.

Output Transformer

Valve (tube) amplifiers almost always use an output transformer to match the high load impedance requirement of the valves (several kilohms) to a low impedance speaker.

Small Signal Transformer

Moving coil phonograph cartridges produce a very small voltage. For this to be amplified with a reasonable signal-noise ratio usually requires a transformer to convert the voltage to the range of the more common moving-magnet cartridges.

Microphones may also be matched to their load with a small transformer, which is mumetal shielded to minimise noise pickup. These transformers are less widely used today, as transistorized buffers are now cheaper.

Interstage and Coupling Transformers

In a push-pull amplifier, an inverted signal is required and can be obtained from a transformer with a center-tapped winding, used to drive two active devices in opposite phase. These phase splitting transformers are not much used today.

Other Types

Transactor

A transactor is a combination of a transformer and a reactor. A transactor has an iron core with an air-gap, which limits the coupling between windings.

Hedgehog

Hedgehog transformers are occasionally encountered in homemade 1920s radios. They are homemade audio interstage coupling transformers.

Enamelled copper wire is wound round the central half of the length of a bundle of insulated iron wire (e.g., florists' wire), to make the windings. The ends of the iron wires are then bent around the electrical winding to complete the magnetic circuit, and the whole is wrapped with tape or string to hold it together.

Variometer and Variocoupler

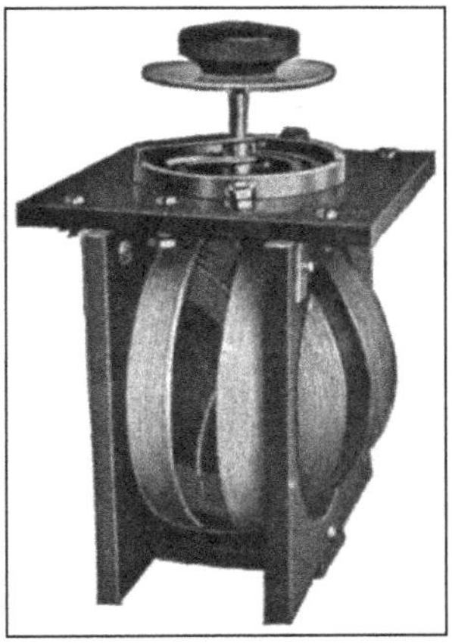

Variometer used in 1920s radio receiver.

A variometer is a type of continuously variable air-core RF inductor with two windings. One common form consisted of a coil wound on a short hollow cylindrical form, with a second smaller coil inside, mounted on a shaft so its magnetic axis can be rotated with respect to the outer coil. The two coils are connected in series. When the two coils are collinear, with their magnetic fields pointed in the same direction, the two magnetic fields add, and the inductance is maximum. If the inner coil is rotated so its axis is at an angle to the outer coil, the magnetic fields do not add and the inductance is less. If the inner coil is rotated so it is collinear with the outer coil but their magnetic fields point in opposite directions, the fields cancel each other out and the inductance is very small or zero. The advantage of the variometer is that inductance can be adjusted continuously, over a wide range. Variometers were widely used in 1920s radio receivers. One of their main uses today is as antenna matching coils to match longwave radio transmitters to their antennas.

The *vario-coupler* was a device with similar construction, but the two coils were not connected but attached to separate circuits. So it functioned as an air-core RF transformer with variable coupling. The inner coil could be rotated from 0° to 90° angle with the outer, reducing the mutual inductance from maximum to near zero.

The pancake coil variometer was another common construction used in both 1920s receivers and transmitters. It consists of two flat spiral coils suspended vertically facing each other, hinged at one side so one could swing away from the other to an angle of 90° to reduce the coupling. The flat spiral design served to reduce parasitic capacitance and losses at radio frequencies.

Pancake or "honeycomb" coil vario-couplers were used in the 1920s in the common Armstrong or "tickler" regenerative radio receivers. One coil was connected to the detector tube's grid circuit. The other coil, the "tickler" coil was connected to the tube's plate (output) circuit. It fed back some of the signal from the plate circuit into the input again, and this positive feedback increased the tube's gain and selectivity.

Rotary Transformer

A rotary (rotatory) transformer is a specialized transformer that couples electrical signals between two parts that rotate in relation to each other—as an alternative to slip rings, which are prone to wear and contact noise. They are commonly used in helical scan magnetic tape applications.

Variable Differential Transformer

A variable differential transformer is a rugged non-contact position sensor. It has two oppositely-phased primaries which nominally produce zero output in the secondary, but any movement of the core changes the coupling to produce a signal.

Resolver and Synchro

The two-phase resolver and related three-phase synchro are rotary position sensors which work over a full 360°. The primary is rotated within two or three secondaries at different angles, and the amplitudes of the secondary signals can be decoded into an angle. Unlike variable differential transformers, the coils, and not just the core, move relative to each other, so slip rings are required to connect the primary.

Resolvers produce in-phase and quadrature components which are useful for computation. Synchros produce three-phase signals which can be connected to other synchros to rotate them in a generator/motor configuration.

Synchro

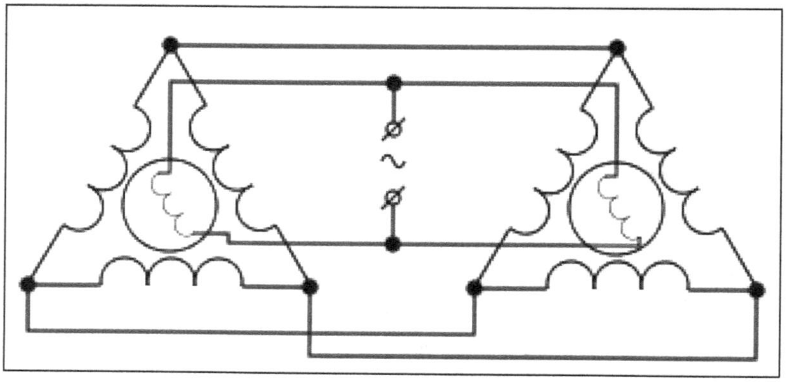

Simple two-synchro system.

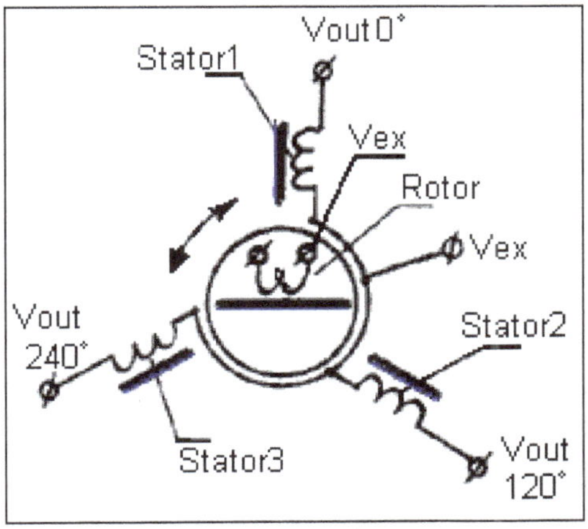

Schematic of a synchro transducer. The complete circle represents the rotor. The solid bars represent the cores of the windings next to them. Power to the rotor is connected by slip rings and brushes, represented by the circles at the ends of the rotor winding. As shown, the rotor induces equal voltages in the 120° and 240° windings, and no voltage in the 0° winding. [Vex] does not necessarily need to be connected to the common lead of the stator star windings.

A synchro is, in effect, a transformer whose primary-to-secondary coupling may be varied by physically changing the relative orientation of the two windings. Synchros are often used for measuring the angle of a rotating machine such as an antenna platform. In its general physical construction, it is much like an electric motor. The primary winding of the transformer, fixed to the rotor, is excited by an alternating current, which by electromagnetic induction, causes currents to flow in three Y-connected secondary windings fixed at 120 degrees to each other on the stator. The relative magnitudes of secondary currents are measured and used to determine the angle of the rotor relative to the stator, or the currents can be used to directly drive a receiver synchro that will rotate in unison with the synchro transmitter. In the latter case, the whole device may be called a selsyn (a portmanteau of *self* and *synchronizing*).

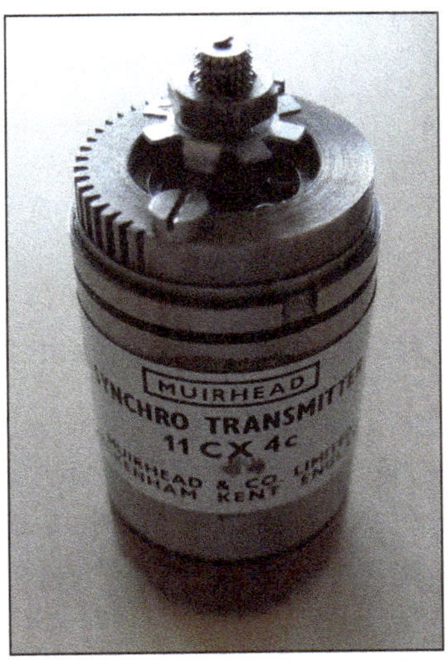

A picture of a synchro transmitter.

Uses

Synchro systems were first used in the control system of the Panama Canal in the early 1900s to transmit lock gate and valve stem positions, and water levels, to the control desks.

View onto the connection description of a synchro transmitter.

Fire-control system designs developed during World War II used synchros extensively, to transmit angular information from guns and sights to an analog fire control computer, and to transmit the desired gun position back to the gun location. Early systems just moved indicator dials, but with the advent of the amplidyne, as well as motor-driven high-powered hydraulic servos, the fire control system could directly control the positions of heavy guns.

Smaller synchros are still used to remotely drive indicator gauges and as rotary position sensors for aircraft control surfaces, where the reliability of these rugged devices is needed. Digital devices such as the rotary encoder have replaced synchros in most other applications.

Selsyn motors were widely used in motion picture equipment to synchronize movie cameras and sound recording equipment, before the advent of crystal oscillators and microelectronics.

Large synchros were used on naval warships, such as destroyers, to operate the steering gear from the wheel on the bridge.

Synchro System Types

There are two types of synchro systems: Torque systems and control systems.

In a torque system, a synchro will provide a low-power mechanical output sufficient to position an indicating device, actuate a sensitive switch or move light loads without power amplification. In simpler terms, a torque synchro system is a system in which the transmitted signal does the usable work. In such a system, accuracy on the order of one degree is attainable.

In a control system, a synchro will provide a voltage for conversion to torque through an amplifier and a servomotor. Control type synchros are used in applications that require large torques or high accuracy such as follow-up links and error detectors in servo, automatic control systems (such as an autopilot system). In simpler terms, a control synchro system is a system in which the transmitted signal controls a source of power which does the usable work.

Quite often, one system will perform both torque and control functions. Individual units are designed for use in either torque or control systems. Some torque units can be used as control units, but control units cannot replace torque units.

Synchro Functional Categories

A synchro will fall into one of eight functional categories. They are as follows:

Torque Transmitter (TX):

> Input: Rotor positioned mechanically or manually by the information to be transmitted.
>
> Output: Electrical output from stator identifying the rotor position supplied to a torque receiver, torque differential transmitter or a torque differential receiver.

Control Transmitter (CX):

> Input: Same as TX.
>
> Output: Electrical output same as TX but supplied to a control transformer or control differential transmitter.

Torque Differential Transmitter (TDX):

> Input: TX output applied to stator; rotor positioned according to amount data from TX that must be modified.
>
> Output: Electrical output from rotor (representing an angle equal to the algebraic sum or difference of rotor position angle and angular data from TX) supplied to torque receivers, another TDX, or a torque differential receiver.

Control Differential Transmitter (CDX):

> Input: Same as TDX but data supplied by CX.
>
> Output: Same as TDX but supplied to only a control transformer or another CDX.

Torque Receiver (TR):

> Input: Electrical angle position data from TX or TDX supplied to stator.
>
> Output: Rotor assumes position determined by electrical input supplied.

Torque Differential Receiver (TDR):

Input: Electrical data supplied from two TX's, two TDX's or from one TX and one TDX (one connected to the rotor and one connected to the stator).

Output: Rotor assumes position equal to the algebraic sum or difference of two angular inputs.

Control Transformer (CT):

Input: Electrical data from CX or CDX applied to stator. Rotor positioned mechanically or manually.

Output: Electrical output from rotor (proportional to sine of the difference between rotor angular position and electrical input angle.

Torque Receiver-Transmitter (TRX):

This synchro was designed as a torque receiver, but may be used as either a transmitter or receiver.

Input: Depending on the application, same as TX.

Output: Depending on the application, same as TX or TR.

Operation

On a practical level, synchros resemble motors, in that there is a rotor, stator, and a shaft. Ordinarily, slip rings and brushes connect the rotor to external power. A synchro transmitter's shaft is rotated by the mechanism that sends information, while the synchro receiver's shaft rotates a dial, or operates a light mechanical load. Single and three-phase units are common in use, and will follow the other's rotation when connected properly. One transmitter can turn several receivers; if torque is a factor, the transmitter must be physically larger to source the additional current. In a motion picture interlock system, a large motor-driven distributor can drive as many as 20 machines, sound dubbers, footage counters, and projectors.

Synchros designed for terrestrial use tend to be driven at 50 or 60 hertz (the mains frequency in most countries), while those for marine or aeronautical use tend to operate at 400 hertz (the frequency of the on-board electrical generator driven by the engines).

Single phase units have five wires: two for an exciter winding (typically line voltage) and three for the output/input. These three are bussed to the other synchros in the system, and provide the power and information to align the shafts of all the receivers. Synchro transmitters and receivers must be powered by the same branch circuit, so to speak; the mains excitation voltage sources must match in voltage and phase. The safest approach is to bus the five or six lines from transmitters and receivers at a common point. Different makes of selsyns, used in interlock systems, have different output voltages. In all cases, three-phase systems will handle more power and operate a bit more smoothly. The excitation is often 208/240 V 3-phase mains power. Many synchros operate on 30 to 60 V AC also.

Synchro transmitters are as described, but 50 and 60-Hz synchro receivers require rotary dampers to keep their shafts from oscillating when not loaded (as with dials) or lightly loaded in high-accuracy applications.

A different type of receiver, called a control transformer (CT), is part of a position servo that includes a servo amplifier and servo motor. The motor is geared to the CT rotor, and when the transmitter's rotor moves, the servo motor turns the CT's rotor and the mechanical load to match the new position. CTs have high-impedance stators and draw much less current than ordinary synchro receivers when not correctly positioned.

Synchro transmitters can also feed synchro to digital converters, which provide a digital representation of the shaft angle.

Synchro Variants

So called 'brushless synchros' use rotary transformers (that have no magnetic interaction with the usual rotor and stator) to feed power to the rotor. These transformers have stationary primaries, and rotating secondaries. The secondary is somewhat like a spool wound with magnet wire, the axis of the spool concentric with the rotor's axis. The "spool" is the secondary winding's core, its flanges are the poles, and its coupling does not vary significantly with rotor position. The primary winding is similar, surrounded by its magnetic core, and its end pieces are like thick washers. The holes in those end pieces align with the rotating secondary poles.

For high accuracy in gun fire control and aerospace work, so called multi-speed synchro data links were used. For instance, a two-speed link had two transmitters, one rotating for one turn over the full range (such as a gun's bearing), while the other rotated one turn for every 10 degrees of bearing. The latter was called a 36-speed synchro. Of course, the gear trains were made accordingly. At the receiver, the magnitude of the 1X channel's error determined whether the "fast" channel was to be used instead. A small 1X error meant that the 36x channel's data was unambiguous. Once the receiver servo settled, the fine channel normally retained control.

For very critical applications, three-speed synchro systems have been used.

So called multispeed synchros have stators with many poles, so that their output voltages go through several cycles for one physical revolution. For two-speed systems, these do not require gearing between the shafts.

Differential synchros are another category. They have three-lead rotors and stators like the stator described above, and can be transmitters or receivers. A differential transmitter is connected between a synchro transmitter and a receiver, and its shaft's position adds to (or subtracts from, depending upon definition) the angle defined by the transmitter. A differential receiver is connected between two transmitters, and shows the sum (or difference, again as defined) between the shaft positions of the two transmitters. There are synchro-like devices called transolvers, somewhat like differential synchros, but with three-lead rotors and four-lead stators.

A resolver is similar to a synchro, but has a stator with four leads, the windings being 90 degrees apart physically instead of 120 degrees. Its rotor might be synchro-like, or have two sets of windings 90 degrees apart. Although a pair of resolvers could theoretically operate like a pair of synchros, resolvers are used for computation.

A special T-connected transformer arrangement invented by Scott ("Scott T") interfaces between resolver and synchro data formats; it was invented to interconnect two-phase AC power with three-phase power, but can also be used for precision applications.

Doubly-fed Electric Machine

Doubly-fed electric machines are electric motors or electric generators where both the field magnet windings and armature windings are separately connected to equipment outside the machine. By feeding adjustable frequency AC power to the field windings, the magnetic field can be made to rotate, allowing variation in motor or generator speed. This is useful, for instance, for generators used in wind turbines.

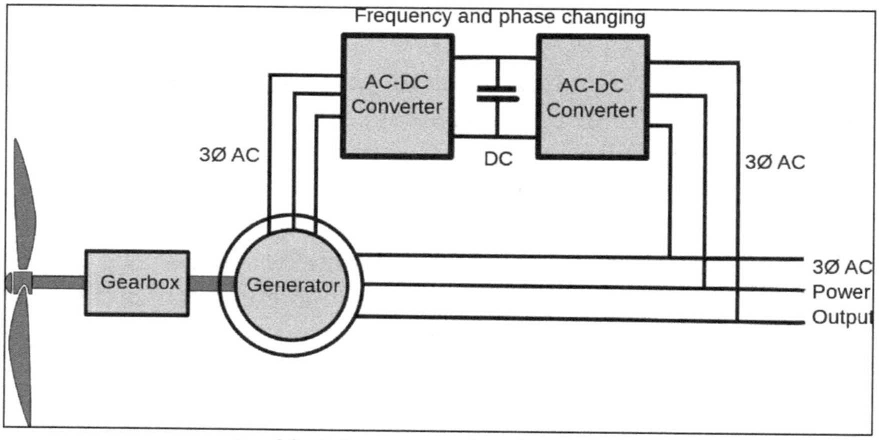

Doubly-fed generator for wind turbine

Doubly fed electrical generators are similar to AC electrical generators, but have additional features which allow them to run at speeds slightly above or below their natural synchronous speed. This is useful for large variable speed wind turbines, because wind speed can change suddenly. When a gust of wind hits a wind turbine, the blades try to speed up, but a synchronous generator is locked to the speed of the power grid and cannot speed up. So large forces are developed in the hub, gearbox, and generator as the power grid pushes back. This causes wear and damage to the mechanism. If the turbine is allowed to speed up immediately when hit by a wind gust, the stresses are lower and the power from the wind gust is converted to useful electricity.

One approach to allowing wind turbine speed to vary is to accept whatever frequency the generator produces, convert it to DC, and then convert it to AC at the desired output frequency using an inverter. This is common for small house and farm wind turbines. But the inverters required for megawatt-scale wind turbines are large and expensive.

Doubly fed generators are one solution to this problem. Instead of the usual field winding fed with DC, and an armature winding where the generated electricity comes out, there are two three-phase windings, one stationary and one rotating, both separately connected to equipment outside the generator. Thus the term "doubly fed".

One winding is directly connected to the output, and produces 3-phase AC power at the desired grid frequency. The other winding (traditionally called the field, but here both windings can be outputs) is connected to 3-phase AC power at variable frequency. This input power is adjusted in frequency and phase to compensate for changes in speed of the turbine.

Adjusting the frequency and phase requires an AC to DC to AC converter. This is usually constructed from very large IGBT semiconductors. The converter is bidirectional, and can pass power in either direction. Power can flow from this winding as well as from the output winding.

History

With its origins in wound rotor induction motors with multiphase winding sets on the rotor and stator, respectively, that was invented by Nikola Tesla in 1888, the rotor winding set of the doubly-fed electric machine is connected to a selection of resistors via multiphase slip rings for starting. However, the slip power was lost in the resistors. Thus means to increase the efficiency in variable speed operation by recovering the slip power were developed. In Krämer (or Kraemer) drives the rotor was connected to an AC and DC machine set that fed a DC machine connected to the shaft of the slip ring machine. Thus the slip power was returned as mechanical power and the drive could be controlled by the excitation currents of the DC machines. The drawback of the Krämer drive is that the machines need to be overdimensioned in order to cope with the extra circulating power. This drawback was corrected in the Scherbius drive where the slip power is fed back to the AC grid by motor generator sets.

The rotating machinery used for the rotor supply was heavy and expensive. Improvement in this respect was the static Scherbius drive where the rotor was connected to a rectifier-inverter set constructed first by mercury arc-based devices and later on with semiconductor diodes and thyristors. In the schemes using a rectifier the power flow was possible only out of the rotor because of the uncontrolled rectifier. Moreover, only sub-synchronous operation as a motor was possible.

Another concept using static frequency converter had a *cycloconverter* connected between the rotor and the AC grid. The cycloconverter can feed power in both directions and thus the machine can be run both sub- and oversynchronous speeds. Large cycloconverter controlled doubly-fed machines have been used to run single phase generators feeding 16 2/3 Hz railway grid in Europe and run the turbines in pumped storage plants.

Today the frequency changer used in applications up to few tens of megawatts consists of two back to back connected *IGBT* inverters.

Several brushless concepts have also been developed in order to get rid of the slip rings that require maintenance.

Double Fed Induction Generator

DFIG for Double Fed Induction Generator, a generating principle widely used in wind turbines. It is based on an induction generator with a multiphase wound rotor and a multiphase slip ring assembly with brushes for access to the rotor windings. It is possible to avoid the multiphase slip ring assembly, but there are problems with efficiency, cost and size. A better alternative is a brushless wound-rotor doubly-fed electric machine.

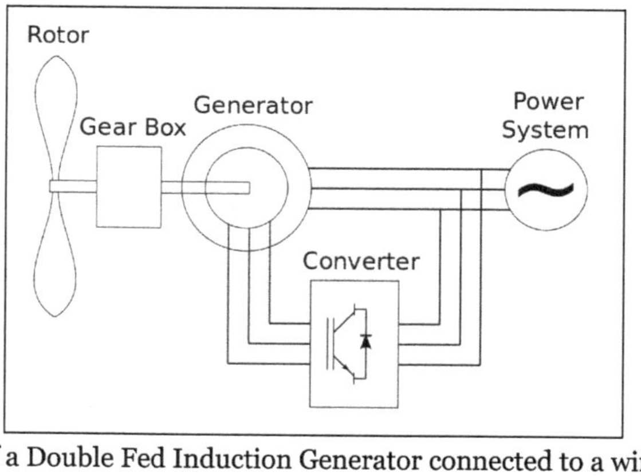

Principle of a Double Fed Induction Generator connected to a wind turbine.

The principle of the DFIG is that rotor windings are connected to the grid via slip rings and back-to-back voltage source converter that controls both the rotor and the grid currents. Thus rotor frequency can freely differ from the grid frequency (50 or 60 Hz). By using the converter to control the rotor currents, it is possible to adjust the active and reactive power fed to the grid from the stator independently of the generator's turning speed. The control principle used is either the two-axis current vector control or direct torque control (DTC). DTC has turned out to have better stability than current vector control especially when high reactive currents are required from the generator.

The doubly-fed generator rotors are typically wound with 2 to 3 times the number of turns of the stator. This means that the rotor voltages will be higher and currents respectively lower. Thus in the typical ± 30% operational speed range around the synchronous speed, the rated current of the converter is accordingly lower which leads to a lower cost of the converter. The drawback is that controlled operation outside the operational speed range is impossible because of the higher than rated rotor voltage. Further, the voltage transients due to the grid disturbances (three- and two-phase voltage dips, especially) will also be magnified. In order to prevent high rotor voltages - and high currents resulting from these voltages - from destroying the IGBTs and diodes of the converter, a protection circuit (called crowbar) is used.

The crowbar will short-circuit the rotor windings through a small resistance when excessive currents or voltages are detected. In order to be able to continue the operation as quickly as possible an active crowbar has to be used. The active crowbar can remove the rotor short in a controlled way and thus the rotor side converter can be started only after 20-60 ms from the start of the grid disturbance when the remaining voltage stays above 15% of the nominal voltage. Thus it is possible to generate reactive current to the grid during the rest of the voltage dip and in this way help the grid to recover from the fault. For zero voltage ride through it is common to wait until the dip ends because with zero voltage it is not possible to know the phase angle where the reactive current should be injected.

As a summary, a doubly-fed induction machine is a wound-rotor doubly-fed electric machine and has several advantages over a conventional induction machine in wind power applications. First, as the rotor circuit is controlled by a power electronics converter, the induction generator is able to both import and export reactive power. This has important consequences for power system stability and allows the machine to support the grid during severe voltage disturbances (low voltage ride through, LVRT). Second, the control of the rotor voltages and currents enables the induction machine to remain synchronized with the grid while the wind turbine speed varies. A variable speed wind turbine utilizes the available wind resource more efficiently than a fixed speed wind turbine, especially during light wind conditions. Third, the cost of the converter is low when compared with other variable speed solutions because only a fraction of the mechanical power, typically 25-30%, is fed to the grid through the converter, the rest being fed to grid directly from the stator. The efficiency of the DFIG is very good for the same reason.

References

- Vučković, Vladan (November 2006). "Interpretation of a Discovery" (PDF). The Serbian Journal of Electrical Engineers. 3 (2). Retrieved 10 February 2013
- Neidhöfer, Gerhard (2007). "Early Three-Phase Power (History)". IEEE Power and Energy Magazine. IEEE Power & Energy Society. 5 (5): 88–100. ISSN 1540-7977. doi:10.1109/MPE.2007.904752

- Hubbell, M.W. (2011). The Fundamentals of Nuclear Power Generation Questions & Answers. Authorhouse. p. 27. ISBN 978-1463424411
- Historical Encyclopedia of Natural and Mathematical Sciences, Volume 1. Springer. 2009-03-06. ISBN 9783540688310
- "Bath County Pumped Storage Station". Dominion Resources, Inc. 2007. Archived from the original on April 4, 2007. Retrieved 2007-03-30
- U.S. DOE (2008). "Improving Motor and Drive System Performance: A Sourcebook for Industry" (PDF). p. 27. Retrieved 31 December 2012

5

Units for Measurement

In electrical engineering, an exact indication of an amount of object or entity is crucial for a successful experiment or design of a system. This chapter's main focus is on the various units of measurement employed in electrical engineering like volt, ampere, coulomb, ohm, farad and henry and they have been comprehensively covered to provide the reader with a clear understanding.

Volt

The volt (symbol: V) is the derived unit for electric potential, electric potential difference (voltage), and electromotive force. The volt is named in honour of the Italian physicist Alessandro Volta (1745–1827), who invented the voltaic pile, possibly the first chemical battery.

Description

The volt is a measure of electric potential. Electrical potential is a type of potential energy, and refers to the energy that could be released if electric current is allowed to flow. An analogy is that a suspended object (a mass) that is said to have gravitational potential energy (as a result of a gravitational field), which is the amount of energy that would be released if the object was allowed to fall.

In alternating current, the voltages increase, decrease and change direction at regular intervals. As a result, voltage for alternating current almost never refers to the voltage at a particular instant, but instead is the root mean square (RMS) voltage, which is a way of defining an effective voltage when calculating power (RMS voltage × RMS current). In most cases, the fact that a voltage is an RMS voltage is not explicitly stated, but assumed.

Definition

One volt is defined as the difference in electric potential between two points of a conducting wire when an electric current of one ampere dissipates one watt of power between those points. It is also equal to the potential difference between two parallel, infinite planes spaced 1 meter apart that create an electric field of 1 newton per coulomb. Additionally, it is the potential difference between two points that will impart one joule of energy per coulomb of charge that passes through it. It can be expressed in terms of SI base units (m, kg, s, and A) as:

$$V = \frac{\text{potential energy}}{\text{charge}} = \frac{N \cdot m}{\text{coulomb}} = \frac{kg \cdot m \cdot m}{s^2 \cdot A \cdot s} = \frac{kg \cdot m^2}{A \cdot s^3}.$$

It can also be expressed as amperes times ohms (current times resistance, Ohm's law), watts per ampere (power per unit current, Joule's law), or joules per coulomb (energy per unit charge), which is also equivalent to electron-volts per elementary charge:

$$V = A \cdot \Omega = \frac{W}{A} = \frac{J}{C} = \frac{eV}{e}.$$

Josephson Junction Definition

The "conventional" volt, V_{90}, defined in 1988 by the 18th General Conference on Weights and Measures and in use from 1990, is implemented using the Josephson effect for exact frequency-to-voltage conversion, combined with the caesium frequency standard. For the Josephson constant, $K_J = 2e/h$ (where e is the elementary charge and h is the Planck constant), the "conventional" value $K_{J\text{-}90}$ is used:

$$K_{J\text{-}90} = 0.4835979 \frac{\text{GHz}}{\mu\text{V}}.$$

This standard is typically realized using a series-connected array of several thousand or tens of thousands of junctions, excited by microwave signals between 10 and 80 GHz (depending on the array design). Empirically, several experiments have shown that the method is independent of device design, material, measurement setup, etc., and no correction terms are required in a practical implementation.

Water-flow Analogy

In the *water-flow analogy* sometimes used to explain electric circuits by comparing them with water-filled pipes, voltage (difference in electric potential) is likened to difference in water pressure.

The relationship between voltage and current is defined (in ohmic devices like resistors) by Ohm's Law. Ohm's Law is analogous to the Hagen–Poiseuille equation, as both are linear models relating flux and potential in their respective systems.

Common Voltages

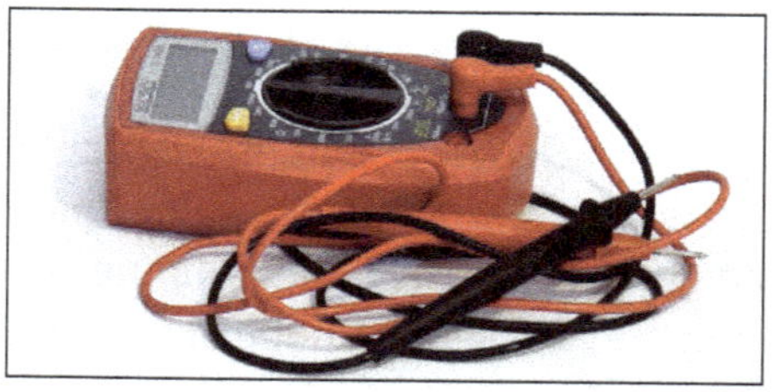

A multimeter can be used to measure the voltage between two positions.

The voltage produced by each electrochemical cell in a battery is determined by the chemistry of that cell. Cells can be combined in series for multiples of that voltage, or additional circuitry added to adjust the voltage to a different level. Mechanical generators can usually be constructed to any voltage in a range of feasibility.

1.5 V C-cell batteries

Nominal voltages of familiar sources:

- Nerve cell resting potential: around −75 mV.
- Single-cell, rechargeable NiMH or NiCd battery: 1.2 V.
- Single-cell, non-rechargeable (e.g., AAA, AA, C and D cells): alkaline battery: 1.5 V; zinc-carbon battery: 1.56 V if fresh and unused.
- $LiFePO_4$ rechargeable battery: 3.3 V.
- Lithium polymer rechargeable battery: 3.75 V.
- Transistor-transistor logic/CMOS (TTL) power supply: 5 V.
- USB: 5 V DC.
- PP3 battery: 9 V.
- Automobile electrical system: nominal 12 V, about 11.8 V discharged, 12.8 V charged, and 13.8–14.4 V while charging (vehicle running).
- Trucks/lorries: 24 V DC (Europe), 12 V DC (North America).
- Household mains electricity AC:
    - 100 V in Japan,
    - 120 V in North America,
    - 230 V in Europe, Asia and Africa,
- Rapid transit third rail: 600–750 V.
- High-speed train overhead power lines: 25 kV at 50 Hz.
- High-voltage electric power transmission lines: 110 kV and up (1.15 MV was the record as of 2005).
- Lightning: Varies greatly, often around 100 MV.

History

In 1800, as the result of a professional disagreement over the galvanic response advocated by Luigi Galvani, Alessandro Volta developed the so-called voltaic pile, a forerunner of the battery, which produced a steady electric current. Volta had determined that the most effective pair of dissimilar metals to produce electricity was zinc and silver. In the 1880s, the International Electrical Congress, now the International Electrotechnical Commission (IEC), approved the volt as the unit for electromotive force. They made the volt equal to 10^8 cgs units of voltage, the cgs system at the time being the customary system of units in science. They chose such a ratio because the cgs unit of voltage is inconveniently small and one volt in this definition is approximately the emf of a Daniell cell, the standard source of voltage in the telegraph systems of the day. At that time, the volt was defined as the potential difference [i.e., what is nowadays called the "voltage (difference)"] across a conductor when a current of one ampere dissipates one watt of power.

The international volt was defined in 1893 as 1/1.434 of the emf of a Clark cell. This definition was abandoned in 1908 in favor of a definition based on the international ohm and international ampere until the entire set of "reproducible units" was abandoned in 1948.

Prior to the development of the Josephson junction voltage standard, the volt was maintained in national laboratories using specially constructed batteries called standard cells. The United States used a design called the Weston cell from 1905 to 1972.

This SI unit is named after Alessandro Volta. As with every International System of Units (SI) unit named for a person, the first letter of its symbol is upper case (V). However, when an SI unit is spelled out in English, it should always begin with a lower case letter (*volt*)—except in a situation where *any* word in that position would be capitalized, such as at the beginning of a sentence or in material using title case. Note that "degree Celsius" conforms to this rule because the "d" is lowercase.— Based on *The International System of Units*, section 5.2.

Ampere

The ampere (SI unit symbol: A), often shortened to "amp", is the SI unit of electric current (dimension symbol: I) and is one of the seven SI base units. It is named after André-Marie Ampère (1775–1836), French mathematician and physicist, considered the father of electrodynamics.

The ampere is equivalent to one coulomb (roughly 7018624100000000000♠6.241×10[18] times the elementary charge) per second. Amperes are used to express flow rate of electric charge. For any point experiencing a current, if the number of charged particles passing through it — or the charge on the particles passing through it — is increased, the amperes of current at that point will proportionately increase.

The ampere should not be confused with the coulomb (also called "ampere-second") or the ampere-hour (A·h). The ampere is a unit of current, the amount of charge transiting per unit time, and the coulomb is a unit of charge. When SI units are used, constant, instantaneous and average current are expressed in amperes (as in "the charging current is 1.2 A") and the charge accumulated, or passed through a circuit over a period of time is expressed in coulombs (as in "the battery charge is 7004300000000000000♠30000 C"). The relation of the ampere (C/s) to the coulomb is the same as that of the watt (J/s) to the joule.

Definition

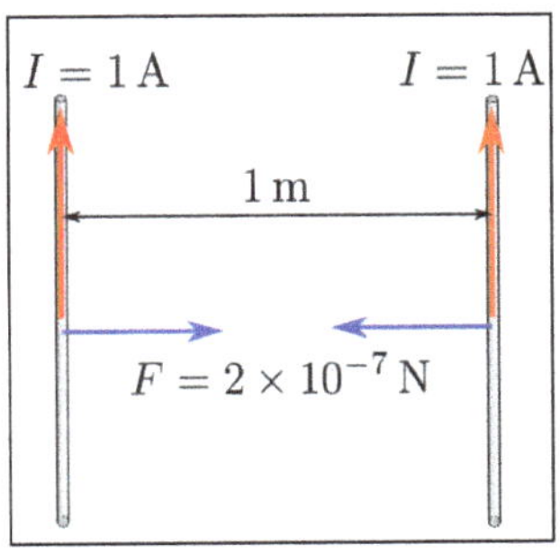

Illustration of the definition of the ampere unit.

Ampère's force law states that there is an attractive or repulsive force between two parallel wires carrying an electric current. This force is used in the formal definition of the ampere, which states that the ampere is the constant current that will produce an attractive force of 2×10^{-7} newtons per metre of length between two straight, parallel conductors of infinite length and negligible circular cross section placed one metre apart in a vacuum.

The SI unit of charge, the coulomb, "is the quantity of electricity carried in 1 second by a current of 1 ampere". Conversely, a current of one ampere is one coulomb of charge going past a given point per second:

$$1\,\mathrm{A} = 1\frac{\mathrm{C}}{\mathrm{s}}.$$

In general, charge Q is determined by steady current I flowing for a time t as $Q = It$.

History

The ampere was originally defined as one tenth of the unit of electric current in the centimetre–gram–second system of units; that unit, now known as the abampere, was defined as the amount of current that generates a force of two dynes per centimetre of length between two wires one centimetre apart. The size of the unit was chosen so that the units derived from it in the MKSA system would be conveniently sized.

The "international ampere" was an early realization of the ampere, defined as the current that would deposit 0.001118000 grams of silver per second from a silver nitrate solution. Later, more accurate measurements revealed that this current is 0.99985 A.

Realization

The standard ampere is most accurately realized using a watt balance, but is in practice maintained via Ohm's law from the units of electromotive force and resistance, the volt and the ohm, since the latter two can be tied to physical phenomena that are relatively easy to reproduce, the Josephson junction and the quantum Hall effect, respectively.

At present, techniques to establish the realization of an ampere have a relative uncertainty of approximately a few parts in 10^7, and involve realizations of the watt, the ohm and the volt.

Proposed Future Definition

Rather than a definition in terms of the force between two current-carrying wires, it has been proposed to define the ampere in terms of the rate of flow of elementary charges. Since a coulomb is approximately equal to 6.2415093×10^{18} elementary charges (such as the those carried by protons, or the negative of those carried by electrons), one ampere is approximately equivalent to 6.2415093×10^{18} elementary charges moving past a boundary in one second. (6.2415093×10^{18} is the reciprocal of the value of the elementary charges in coulombs.) The proposed change would define 1 A as being the current in the direction of flow of a particular number of elementary charges per second. In 2005, the International Committee for Weights and Measures (CIPM) agreed to study the proposed change. The new definition was discussed at the 25th General Conference on Weights and Measures (CGPM) in 2014 but for the time being was not adopted.

Everyday Examples

The current drawn by typical constant-voltage energy distribution systems is usually dictated by the power (watt) consumed by the system and the operating voltage. For this reason the examples given below are grouped by voltage level.

Portable Devices

- Hearing aid (typically 1 mW at 1.4 V): 0.7 mA.

Internal Combustion Engine Vehicles – 12 V DC

A typical motor vehicle has a 12 V battery. The various accessories that are powered by the battery might include:

- Instrument panel light (typically 2 W): 166 mA.
- Headlights (typically 60 W): 5 A each.
- Starter motor (typically 1–2 kW): 80–160 A

North American Domestic Supply – 120 V AC

Most Canada, Mexico and United States domestic power suppliers run at 120 V.

Household circuit breakers typically provide a maximum of 15 A or 20 A of current to a given set of outlets:

- 22-inch/56-centimeter portable television (35 W): 290 mA.
- Tungsten light bulb (60–100 W): 500–830 mA.
- Toaster, kettle (1.5 kW): 12.5 A.
- Hair dryer (1.8 kW): 15 A.

European & Commonwealth Domestic Supply – 230-240 V AC

Most European domestic power supplies run at 230 V, and most Commonwealth domestic power supplies run at 240 V. For the same amount of power (in watts), the current drawn by a particular European or Commonwealth appliance (in Europe or a Commonwealth country) will be less than for an equivalent North American appliance. Typical circuit breakers will provide 16 A.

The current drawn by a number of typical appliances are:

- 22-inch/56-centimeter portable television (35 W): 145–150 mA.
- Tungsten light bulb (60–100 W): 240–450 mA.
- Compact fluorescent lamp (11–30 W): 56–112 mA.
- Toaster, kettle (2 kW): 9 A.
- Immersion heater (4.6 kW): 19–20 A.

Coulomb

The coulomb (unit symbol: **C**) is the International System of Units (SI) unit of electric charge. It is the charge (symbol: *Q* or *q*) transported by a constant current of one ampere in one second:

$$1\mathrm{C} = 1\mathrm{A} \cdot 1\mathrm{s}$$

Thus, it is also the amount of excess charge on a capacitor of one farad charged to a potential difference of one volt:

$$1\mathrm{C} = 1\mathrm{F} \cdot 1\mathrm{V}$$

It is equivalent to the charge of approximately 6.242×10^{18} (1.036×10^{-5} mol) protons, and -1 C is equivalent to the charge of approximately 6.242×10^{18} electrons.

Name and Notation

This SI unit is named after Charles-Augustin de Coulomb. As with every International System of Units (SI) unit named for a person, the first letter of its symbol is upper case (C). However, when an SI unit is spelled out in English, it should always begin with a lower case letter (*coulomb*)—except in a situation where *any* word in that position would be capitalized, such as at the beginning of a sentence or in material using title case. Note that "degree Celsius" conforms to this rule because the "d" is lowercase.— Based on *The International System of Units*, section 5.2.

Definition

The SI system defines the coulomb in terms of the ampere and second: 1 C = 1 A × 1 s. The second is defined in terms of a frequency naturally emitted by caesium atoms. The ampere is defined using Ampère's force law; the definition relies in part on the mass of the international prototype kilogram, a metal cylinder housed in France. In practice, the watt balance is used to measure amperes with the highest possible accuracy.

Since the charge of one electron is known to be about (98)×10^{-19} C, −1 C can also be considered the charge of roughly 6.241509×10^^{18} electrons (or +1 C the charge of that many positrons or protons), where the number is the reciprocal of 1.602177×10^$^{-19}$.

The proposed redefinition of the ampere and other SI base units would have the effect of fixing the numerical value of the fundamental charge to an explicit constant expressed in coulombs, and therefore it would implicitly fix the value of the coulomb when expressed as a multiple of the fundamental charge (the numerical values of those quantities are the multiplicative inverses of each other).

SI Prefixes

SI multiples for coulomb (C)					
Submultiples			**Multiples**		
Value	**SI symbol**	**Name**	**Value**	**SI symbol**	**Name**
10^{-1} C	dC	decicoulomb	10^{1} C	daC	decacoulomb
10^{-2} C	cC	centicoulomb	10^{2} C	hC	hectocoulomb
10^{-3} C	**mC**	**millicoulomb**	10^{3} C	kC	kilocoulomb
10^{-6} C	**µC**	**microcoulomb**	10^{6} C	MC	megacoulomb
10^{-9} C	**nC**	**nanocoulomb**	10^{9} C	GC	gigacoulomb
10^{-12} C	**pC**	**picocoulomb**	10^{12} C	TC	teracoulomb
10^{-15} C	fC	femtocoulomb	10^{15} C	PC	petacoulomb
10^{-18} C	aC	attocoulomb	10^{18} C	EC	exacoulomb
10^{-21} C	zC	zeptocoulomb	10^{21} C	ZC	zettacoulomb
10^{-24} C	yC	yoctocoulomb	10^{24} C	YC	yottacoulomb
Common multiples are in bold face.					

Conversions

- One coulomb is the magnitude (absolute value) of electrical charge in 6.24150934(14)×10^18 protons or electrons.
- The inverse of this number gives the elementary charge of 1.6021766208 (98)×10^{-19} C.
- The magnitude of the electrical charge of one mole of elementary charges (approximately 6.022×10^{23}, or Avogadro's number) is known as a faraday unit of charge (closely related to the Faraday constant). One faraday equals 96485.3399 coulombs. In terms of Avogadro's number (N_A), one coulomb is equal to approximately 1.036 × N_A×10^{-5} elementary charges.
- One ampere-hour = 3600 C, 1 mA·h = 3.6 C.
- One statcoulomb (statC), the obsolete CGS electrostatic unit of charge (esu), is approximately 3.3356×10^{-10} C or about one-third of a nanocoulomb.

Relation to Elementary Charge

The elementary charge, the charge of a proton (equivalently, the negative of the charge of an electron), is approximately 1.6021766208(98)×10^{-19} C. In SI, the elementary charge in coulombs is an approximate value: no experiment can be infinitely accurate. However, in other unit systems, the elementary charge has an *exact* value by definition, and other charges are ultimately measured relative to the elementary charge. For example, in conventional electrical units, the values of the Josephson constant K_J and von Klitzing constant R_K are exact defined values (written $K_{J\text{-}90}$ and $R_{K\text{-}90}$), and it follows that the elementary charge $e = 2/(K_J R_K)$ is also an exact defined value in this unit system. Specifically, e_{90} = (2×10^{-9})/(25812.807 × 483597.9) C exactly. SI itself may someday change its definitions in a similar way. For example, one possible proposed redefinition is "the ampere...is [defined] such that the value of the elementary charge *e* (charge on a proton) is exactly 1.602176487×10^{-19} coulombs", (in which the numeric value is the 2006 CODATA recommended value, since superseded). This proposal is not yet accepted as part of the SI; the SI definitions are unlikely to change until at least 2015.

In Everyday Terms

- The charges in static electricity from rubbing materials together are typically a few microcoulombs.
- The amount of charge that travels through a lightning bolt is typically around 15 C, although large bolts can be up to 350 C.
- The amount of charge that travels through a typical alkaline AA battery from being fully charged to discharged is about 5 kC = 5000 C ≈ 1400 mA·h.
- According to Coulomb's law, two negative point charges of −1 C, placed one meter apart, would experience a repulsive force of 9×10^9 N, a force roughly equal to the weight of 920000 metric tons of mass on the surface of the Earth.

- The hydraulic analogy uses everyday terms to illustrate movement of charge and the transfer of energy. The analogy equates charge to a volume of water, and voltage to pressure. One coulomb equals (the negative of) the charge of 6.24×10^{18} elec-trons. The amount of energy transferred by the flow of 1 coulomb can vary; for example, 300 times fewer electrons flow through a lightning bolt than in the discharge of an AA bat-tery, but the total energy transferred by the flow of the lightning's electrons is 300 million times greater.

Ohm

The ohm (symbol: Ω) is the SI derived unit of electrical resistance, named after German physicist Georg Simon Ohm. Although several empirically derived standard units for expressing electrical resistance were developed in connection with early telegraphy practice, the British Association for the Advancement of Science proposed a unit derived from existing units of mass, length and time and of a convenient size for practical work as early as 1861. The definition of the ohm was revised several times. Today the definition of the ohm is expressed from the quantum Hall effect.

Definition

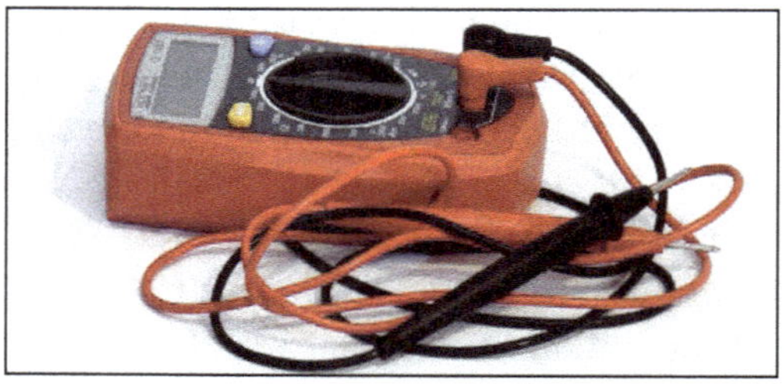

A multimeter can be used to measure resistance in ohms, among other things.

The ohm is defined as an electrical resistance between two points of a conductor when a constant potential difference of 1 volt, applied to these points, produces in the conductor a current of 1 ampere, the conductor not being the seat of any electromotive force.

$$\Omega = \frac{V}{A} = \frac{1}{S} = \frac{W}{A^2} = \frac{V^2}{W} = \frac{s}{F} = \frac{J \cdot s}{C^2} = \frac{kg \cdot m^2}{s \cdot C^2} = \frac{J}{s \cdot A^2} = \frac{kg \cdot m^2}{s^3 \cdot A^2}$$

in which the following units appear: volt (V), ampere (A), siemens (S), watt (W), second (s), farad (F), joule (J), kilogram (kg), metre (m), and coulomb (C).

In many cases the resistance of a conductor in ohms is approximately constant within a certain range of voltages, temperatures, and other parameters. These are called linear resistors. In other cases resistance varies (e.g., thermistors).

Commonly used multiples and submultiples in electrical and electronic usage are the microohm, milliohm, kilohm, megohm, and gigaohm; the term "gigohm", though not official, is in common use for the latter.

In alternating current circuits, electrical impedance is also measured in ohms (Ω).

Conversions

The siemens (symbol: S) is the SI derived unit of electric conductance and admittance, also known as the mho (ohm spelled backwards, symbol is ℧); it is the reciprocal of resistance in ohms (Ω).

Power as a Function of Resistance

The power dissipated by a resistor may be calculated from its resistance, and the voltage or current involved. The formula is a combination of Ohm's Law and Joule's law:

$$P = V \cdot I = \frac{V^2}{R} = I^2 \cdot R$$

where:

P = power in watts.

R = resistance in ohms.

V = voltage across the resistor.

I = current through the resistor in amps.

A linear resistor has a constant resistance value over all applied voltages or currents; many practical resistors are linear over a useful range of currents. Non-linear resistors have a value that may vary depending on the applied voltage (or current). Where alternating current is applied to the circuit,(or where the resistance value is a function of time), the relation above is true at any instant but calculation of average power over an interval of time will require integration of "instantaneous" power over that interval.

History

The rapid rise of electrotechnology in the last half of the 19th century created a demand for a rational, coherent, consistent, and international system of units for electrical quantities. Telegraphers and other early uses of electricity in the 19th century needed a practical standard unit of measurement for resistance. Resistance was often expressed as a multiple of the resistance of a standard length of telegraph wires; different agencies used different bases for a standard, so units were not readily interchangeable. Electrical units so defined were not a coherent system with the units for energy, mass, length, and time, requiring conversion factors to be used in calculations relating energy or power to resistance.

Two different methods of establishing a system of electrical units can be chosen. Various artifacts, such as a length of wire or a standard electrochemical cell, could be specified as producing defined quantities for resistance, voltage, and so on. Alternatively, the electrical units can be related to the mechanical units by defining, for example, a unit of current that gives a specified force between two wires, or a unit of charge that gives a unit of force between two unit charges. This latter method ensures coherence with the units of energy. Defining a unit for resistance that is coherent with units of energy and time in effect also requires defining units for potential and current. It is desirable that one unit of electrical potential will force one unit of electric current through one unit of electrical resistance, doing one unit of work in one unit of time, otherwise all electrical calculations will require conversion factors.

Since so-called "absolute" units of charge and current are expressed as combinations of units of mass, length, and time, dimensional analysis of the relations between potential, current, and resistance show that resistance is expressed in units of length per time — a velocity. Some early definitions of a unit of resistance, for example, defined a unit resistance as one quadrant of the Earth per second.

The absolute-units system related magnetic and electrostatic quantities to metric base units of mass, time, and length. These units had the great advantage of simplifying the equations used in the solution of electromagnetic problems, and eliminated conversion factors in calculations about electrical quantities. However, the CGS units turned out to have impractical sizes for practical measurements.

Various artifact standards were proposed as the definition of the unit of resistance. In 1860 Werner Siemens published a suggestion for a reproducible resistance standard in Poggendorffs Annalen der Physik und Chemie. He proposed a column of pure mercury, of one square millimetre cross section, one metre long: Siemens mercury unit. However, this unit was not coherent with other units. One proposal was to devise a unit based on a mercury column that would be coherent - in effect, adjusting the length to make the resistance one ohm. Not all users of units had the resources to carry out metrology experiments to the required precision, so working standards notionally based on the physical definition were required.

In 1861, Latimer Clark and Sir Charles Bright presented a paper at the British Association for the Advancement of Science meeting suggesting that standards for electrical units be established and suggesting names for these units derived from eminent philosophers, 'Ohma', 'Farad' and 'Volt'. The BAAS in 1861 appointed a committee including Maxwell and Thomson to report upon Standards of Electrical Resistance. Their objectives were to devise a unit that was of convenient size, part of a complete system for electrical measurements, coherent with the units for energy, stable, reproducible and based on the French metrical system. In the third report of the committee, 1864, the resistance unit is referred to as "B.A. unit, or Ohmad". By 1867 the unit is referred to as simply Ohm.

The B.A. ohm was intended to be 10^9 CGS units but owing to an error in calculations the definition was 1.3% too small. The error was significant for preparation of working standards.

On September 21, 1881 the *Congrès internationale d'électriciens* (international conference of electricians) defined a *practical* unit of Ohm for the resistance, based on CGS units, using a mercury column at zero deg. Celsius, similar to the apparatus suggested by Siemens.

A *legal* ohm, a reproducible standard, was defined by the international conference of electricians at Paris in 1884 as the resistance of a mercury column of specified weight and 106 cm long; this was a compromise value between the B. A. unit (equivalent to 104.7 cm), the Siemens unit (100 cm by definition), and the CGS unit. Although called "legal", this standard was not adopted by any national legislation. The "international" ohm was defined as a mercury column 106.3 cm long of mass 14.4521 grams and 0 °C at the International Electrical Conference 1893 in Chicago. This definition became the basis for the legal definition of the ohm in several countries. In 1908, the next Electrical Conference confirmed this definition. The mercury column standard was maintained until the 1948 General Conference on Weights and Measures, at which the ohm was redefined in absolute terms instead of as an artifact standard.

By the end of the 19th century, units were well-understood and consistent. Definitions would change with little effect on commercial uses of the units. Advances in metrology allowed definitions to be formulated with a high degree of precision and repeatability.

Historical Units of Resistance

Unit	Definition	Value in B.A. ohms	Remarks
Absolute foot/second x 10^7	using Imperial units	.3048	considered obsolete even in 1884
Thomson's unit	using Imperial units	.3202	100 million feet/second, considered obsolete even in 1884
Jacobi copper unit	A specified copper wire 25 feet long weighing 345 grains	.6367	Used in 1850s
Weber's absolute unit × 10^7	Based on the metre and the second	0.9191	
Siemens mercury unit	1860. A column of pure mercury	.9537	100 cm and 1 mm² cross section at 0 °C
British Association (B.A.) "ohm"	1863	1.000	Standard coils deposited at Kew Observatory in 1863
Digney, Breguet, Swiss		9.266–10.420	Iron wire 1 km long and 4 square mm cross section
Matthiessen		13.59	One mile of 1/16 inch diameter pure annealed copper wire at 15.5 °C
Varley		25.61	One mile of special 1/16 inch diameter copper wire
German mile		57.44	A German mile (8,238 yard) of iron wire 1/6th inch diameter
Abohm		10^{-9}	Electromagnetic absolute unit in cm-gram-second units
Statohm		$8.987551787 \times 10^{11}$	Electrostatic absolute unit in cm-gram-second units

Realization of Standards

The mercury column method of realizing a physical standard ohm turned out to be difficult to reproduce, owing to the effects of non-constant cross section of the glass tubing. Various resistance coils were constructed by the British Association and others, to serve as physical artifact standards for the unit of resistance. The long-term stability and reproducibility of these artifacts was an ongoing field of research, as the effects of temperature, air pressure, humidity, and time on the standards were detected and analyzed.

Artifact standards are still used, but metrology experiments relating accurately-dimensioned inductors and capacitors provided a more fundamental basis for the definition of the ohm. Since

1990 the quantum Hall effect has been used to define the ohm with high precision and repeatability. The quantum Hall experiments are used to check the stability of working standards that have convenient values for comparison.

Symbol

When preparing electronic documents, some document editing software applications have used the Symbol typeface to render the character Ω. Where the font is not supported, a W is displayed instead ("10 W" instead of "10 Ω", for instance). As W represents the watt, the SI unit of power, this can lead to confusion.

In the electronics industry it is common to use the character R instead of the Ω symbol, thus, a 10 Ω resistor may be represented as 10R. This is the British standard BS 1852 code. It is used in many instances where the value has a decimal place. For example, 5.6 Ω is listed as 5R6. This method avoids overlooking the decimal period, which may not be rendered reliably on components or when duplicating documents.

In documents printed before WWII the unit symbol often consisted of the raised lowercase omega (ω), such that 56 Ω was written as 56^{ω}.

Unicode encodes the symbol as U+2126 Ω OHM SIGN, distinct from Greek omega among letter-like symbols, but it is only included for backwards compatibility and the Greek uppercase omega character U+03A9 Ω GREEK CAPITAL LETTER OMEGA (HTML Ω · Ω) is preferred. In DOS and Windows, the alt code ALT 234 may produce the Ω symbol. In Mac OS, Opt+Z does the same.

Farad

The farad (symbol: F) is the SI derived unit of electrical capacitance, the ability of a body to store an electrical charge. It is named after the English physicist Michael Faraday.

Definition

One farad is defined as the capacitance across which, when charged with one coulomb, there is a potential difference of one volt. Equally, one farad can be described as the capacitance which stores a one coulomb charge across a potential difference of one volt.

The relationship between capacitance, charge and potential difference is linear. For example, if the potential difference across a capacitor is halved, the quantity of charge stored by that capacitor will also be halved.

For most applications, the farad is an impractically large unit of capacitance. Most electrical and electronic applications are covered by the following SI prefixes:

- 1 mF (millifarad, one thousandth (10^{-3}) of a farad) = 1000 μF = 1000000 nF.
- 1 μF (microfarad, one millionth (10^{-6}) of a farad) = 1000 nF = 1000000 pF.

- 1 nF (nanofarad, one billionth (10^{-9}) of a farad) = 1000 pF.
- 1 pF (picofarad, one trillionth (10^{-12}) of a farad).

Equalities

A farad has the base SI representation of: $s^4 \times A^2 \times m^{-2} \times kg^{-1}$

It can further be expressed as:

$$F = \frac{A \cdot s}{V} = \frac{J}{V^2} = \frac{W \cdot s}{V^2} = \frac{C}{V} = \frac{C^2}{J} = \frac{C^2}{N \cdot m} = \frac{s^2 \cdot C^2}{m^2 \cdot kg} = \frac{s^4 \cdot A^2}{m^2 \cdot kg} = \frac{s}{\Omega} = \frac{s^2}{H}$$

where F=farad, A=ampere, V=volt, C=coulomb, J=joule, m=metre, N=newton, s=second, W=watt, kg=kilogram, Ω=ohm, H=henry.

History

The term "farad" was originally coined by Latimer Clark and Charles Bright in 1861, in honor of Michael Faraday, for a unit of quantity of charge but, starting in 1881 at the International Congress of Electricians in Paris, the name farad was officially used for the unit of electrical capacitance.

Explanation

Examples of different types of capacitors.

A capacitor consists of two conducting surfaces, frequently referred to as plates, separated by an insulating layer usually referred to as a dielectric. The original capacitor was the Leyden jar developed in the 18th century. It is the accumulation of electric charge on the plates that results in capacitance. Modern capacitors are constructed using a range of manufacturing techniques and materials to provide the extraordinarily wide range of capacitance values used in electronics applications from femtofarads to farads, with maximum-voltage ratings ranging from a few volts to several kilovolts.

Values of capacitors are usually specified in farads (F), microfarads (μF), nanofarads (nF) and picofarads (pF). The millifarad is rarely used in practice (a capacitance of 4.7 mF (0.0047 F), for example, is instead written as 4700 μF), while the nanofarad is uncommon in North America. The size of commercially available capacitors ranges from around 0.1 pF to 5000F (5 kF) supercapacitors. Parasitic capacitance in high-performance integrated circuits can be measured in femtofarads (1 fF = 0.001 pF = 10^{-15} F), while high-performance test equipment can detect changes in capacitance on the order of tens of attofarads (1 aF = 10^{-18} F).

A value of 0.1 pF is about the smallest available in capacitors for general use in electronic design, since smaller ones would be dominated by the parasitic capacitances of other components, wiring or printed circuit boards. Capacitance values of 1 pF or lower can be achieved by twisting two short lengths of insulated wire together.

The capacitance of the Earth's ionosphere with respect to the ground is calculated to be about 1 F.

Informal and Deprecated Terminology

The picofarad is sometimes colloquially pronounced as "puff" or "pic", as in "a ten-puff capacitor". Similarly, "mic" (pronounced "mike") is sometimes used informally to signify microfarads. If the Greek letter μ is not available, the notation "uF" is often used as a substitute for "μF" in electronics literature. A "micro-microfarad" (μμF, and confusingly often mmf or MMF), an obsolete unit sometimes found in older texts, is the equivalent of a picofarad. In texts prior to 1960, and on capacitor packages even much more recently, mf or MFD rather than the modern μF frequently represented microfarads. Similarly, mmf represented picofarads.

Related Concepts

The reciprocal of capacitance is called electrical elastance, the (non-standard, non-SI) unit of which is the daraf.

CGS Units

The abfarad (abbreviated abF) is an obsolete CGS unit of capacitance equal to 10^9 farads (1 gigafarad, GF).

The statfarad (abbreviated statF) is a rarely used CGS unit equivalent to the capacitance of a capacitor with a charge of 1 statcoulomb across a potential difference of 1 statvolt. It is $1/(10^{-5}c^2)$ farad, approximately 1.1126 picofarads.

Henry

The henry (symbol H) is the unit of electrical inductance in the International System of Units. The unit is named after Joseph Henry (1797–1878), the American scientist who discovered electromagnetic induction independently of and at about the same time as Michael Faraday (1791–1867) in England. The magnetic permeability of vacuum is $4\pi \times 10^{-7}$ H m^{-1} (henry per metre).

The United States National Institute of Standards and Technology recommends English-speaking users of SI to write the plural as *henries*.

Definition

The inductance of an electric circuit is one henry when an electric current that is changing at one ampere per second results in an electromotive force of one volt across the inductor :

$$v(t) = L\frac{di}{dt},$$

where $v(t)$ denotes the resulting voltage across the circuit, $i(t)$ is the current through the circuit, and L is the inductance of the circuit.

The henry is a derived unit based on four of the seven base units of the International System of Units: kilogram (kg), meter (m), second (s), and ampere (A). Expressed in combinations of SI units, the henry is:

$$\mathrm{H} = \frac{\mathrm{kg \cdot m^2}}{\mathrm{s^2 \cdot A^2}} = \frac{\mathrm{kg \cdot m^2}}{\mathrm{C^2}} = \frac{\mathrm{J}}{\mathrm{A^2}} = \frac{\mathrm{T \cdot m^2}}{\mathrm{A}} = \frac{\mathrm{Wb}}{\mathrm{A}} = \frac{\mathrm{V \cdot s}}{\mathrm{A}} = \frac{\mathrm{s^2}}{\mathrm{F}} = \frac{1}{\mathrm{F \cdot Hz^2}} = \Omega \cdot \mathrm{s}$$

where

Wb = weber,
T = tesla,
J = joule,
m = meter,
s = second,
A = ampere,
V = volt,
C = coulomb,
F = farad,
Hz = hertz,
Ω = ohm

As with every International System of Units (SI) unit named for a person, the first letter of its symbol is upper case (H). However, when an SI unit is spelled out in English, it should always begin with a lower case letter (henry)—except in a situation where any word in that position would be capitalized, such as at the beginning of a sentence or in material using title case. Note that "degree Celsius" conforms to this rule because the "d" is lowercase.— Based on The International System of Units , section 5.2.

References

- Serway, Raymond A; Jewett, JW (2006). Serway's principles of physics: a calculus based text (Fourth ed.). Belmont, CA: Thompson Brooks/Cole. p. 746. ISBN 0-53449143-X
- Monk, Paul MS (2004), Physical Chemistry: Understanding our Chemical World, John Wiley & Sons, ISBN 0-471-49180-2
- Hill, Paul Horowitz; Winfield; Winfield, Hill (2015). The Art of Electronics (3. ed.). Cambridge [u.a.]: Cambridge Univ. Press. p. 689. ISBN 978-0-521-809269
- Platt, Charles (2009). Make: Electronics: Learning Through Discovery. O'Reilly Media. p. 61. ISBN 9781449388799. Retrieved 2014-07-22.

- Hunt, Bruce J (1994). "The Ohm Is Where the Art Is: British Telegraph Engineers and the Development of Electrical Standards" (PDF). Osiris. 2nd. 9: 48–63. doi:10.1086/368729. Retrieved 27 February 2014
- Clark, Latimer; Bright, Sir Charles (1861-11-09). "Measurement of Electrical Quantities and Resistance". The Electrician. 1 (1): 3–4. Retrieved 27 February 2014
- "CODATA Value: elementary charge". The NIST Reference on Constants, Units, and Uncertainty. US National Institute of Standards and Technology. June 2015. Retrieved 2015-09-22

Permissions

All chapters in this book are published with permission under the Creative Commons Attribution Share Alike License or equivalent. Every chapter published in this book has been scrutinized by our experts. Their significance has been extensively debated. The topics covered herein carry significant information for a comprehensive understanding. They may even be implemented as practical applications or may be referred to as a beginning point for further studies.

We would like to thank the editorial team for lending their expertise to make the book truly unique. They have played a crucial role in the development of this book. Without their invaluable contributions this book wouldn't have been possible. They have made vital efforts to compile up to date information on the varied aspects of this subject to make this book a valuable addition to the collection of many professionals and students.

This book was conceptualized with the vision of imparting up-to-date and integrated information in this field. To ensure the same, a matchless editorial board was set up. Every individual on the board went through rigorous rounds of assessment to prove their worth. After which they invested a large part of their time researching and compiling the most relevant data for our readers.

The editorial board has been involved in producing this book since its inception. They have spent rigorous hours researching and exploring the diverse topics which have resulted in the successful publishing of this book. They have passed on their knowledge of decades through this book. To expedite this challenging task, the publisher supported the team at every step. A small team of assistant editors was also appointed to further simplify the editing procedure and attain best results for the readers.

Apart from the editorial board, the designing team has also invested a significant amount of their time in understanding the subject and creating the most relevant covers. They scrutinized every image to scout for the most suitable representation of the subject and create an appropriate cover for the book.

The publishing team has been an ardent support to the editorial, designing and production team. Their endless efforts to recruit the best for this project, has resulted in the accomplishment of this book. They are a veteran in the field of academics and their pool of knowledge is as vast as their experience in printing. Their expertise and guidance has proved useful at every step. Their uncompromising quality standards have made this book an exceptional effort. Their encouragement from time to time has been an inspiration for everyone.

The publisher and the editorial board hope that this book will prove to be a valuable piece of knowledge for students, practitioners and scholars across the globe.

Index

www.ingramcontent.com/pod-product-compliance
Lightning Source LLC
LaVergne TN
LVHW080626210926
848485LV00003B/217